THE CONTAINER VICTORY GARDEN

A Beginner's Guide to Growing Your Own Groceries

Maggie Stuckey

Art by

Janice Minjin Yang

and

Lee Johnston

The Container Victory Garden

Published by Harper Celebrate, an imprint of HarperCollins Focus LLC.

All stories used by permission.

Full-color illustrations by Janice Minjin Yang.

Black-and-white illustrations by Lee Johnston.

Photography credits noted next to images except for the following: page iv, Shutterstock/lovelyday12; page vi, Shutterstock/Danita Delimont; page viii, Shutterstock/PhotoIris2021; page x, Shutterstock/MEE KO DONG; page xv, Shutterstock/Red pepper; page xvi, Shutterstock/Stanislavskyi; page 2, Library of Congress/lithograph by the Stecher-Traung Lithograph Corporation; page 15, Library of Congress; page 19, Shutterstock/VICUSCHKA; page 41, Shutterstock/MangoNic; page 45, Shutterstock/DimaBerlin; page 69, Shutterstock/Sunny Forest; page 84, Shutterstock/CobraCZ; page 90, Shutterstock/Shaiith; page 91, Shutterstock/ami mataraj; page 93, Shutterstock/Olga Bondarenko; page 96, Shutterstock/Paul Maguire; page 100, Shutterstock/Nadia Nice; page 102, Shutterstock /nblx; page 103, Shutterstock/ArtCookStudio; page 106, Shutterstock/natalia bulatova; page 112, Shutterstock/eurobanks; page 117, Shutterstock/oksana2010; page 125, Shutterstock/Jean Faucett; page 126, Shutterstock/HandmadePictures; page 131, Shutterstock /Artist1704; page 141, Shutterstock/Shebeko; page 145, Shutterstock/Tatevosian Yana; page 148, Shutterstock/stockcreations; page 149, Shutterstock/Brent Hofacker; page 155, Shutterstock/Aaron of L.A. Photography; page 157, Shutterstock/Bildagentur Zoonar GmbH; page 158, Shutterstock/Gts; page 161, Shutterstock/AnikonaAnn; page 163, Shutterstock/creativeneko; page 164, Shutterstock/VH-studio (top), Shutterstock/Oxana Denezhkina (bottom); page 167, Shutterstock/Maren Winter; page 170, Shutterstock/nnattalli; page 173, Shutterstock/Snowboy; page 179, Shutterstock/Malykalexa; page 180, Shutterstock/Patrick Foto; page 182, Shutterstock/Julia Gardener; page 184, Shutterstock/photosgenius; page 185, Shutterstock/RANGINAPHOTOGRAPHY; page 186, Shutterstock/HiTecherZ; page 189, Shutterstock/aniana; page 195, Shutterstock/photographyfirm; page 197, Shutterstock/oioioi; page 201, Shutterstock/ValdisOsins; page 202, Shutterstock/Kutterlvaserova Stuchelova; page 204, iStock/kali9; page 214, Shutterstock/PeopleImages.com - Yuri A; page 216, Shutterstock/eurobanks; page 218, Shutterstock/stock_studio; pages 219–27, Shutterstock/Nadia Nice (top); pages 228–29, Shutterstock /Brent Hofacker (top); page 230–31, Shutterstock/oioioi (top); page 232, Shutterstock/KRIACHKO OLEKSII; page 245, Shutterstock/Olga Miltsova; page 246, Shutterstock/JUANST; page 253 and 254, Shutterstock/Olga Miltsova; page 256, Shutterstock Oksana Kuznetsova Dnepr.

ISBN-13: 978-0-7852-5581-9 (Audio)
ISBN-13: 978-0-7852-5579-6 (ebook)
ISBN-13: 978-0-7852-5576-5 (SC)

Library of Congress Control Number: 2022941702

Printed in Canada

23 24 25 26 27 TC 10 9 8 7 6 5 4 3 2 1

This book is dedicated to everyone who works in the
horticulture industry. Every single one of you.
In the good times, you provide us with the raw materials for joy, beauty, satisfaction,
solace, and good food. In the bad times, such as we have recently endured, when we
needed that solace more than ever, you carry on. Working under impossible conditions,
facing unheard-of new problems every blessed day, you still manage to get us the raw
materials for joy, and beauty, and peace. And you do it with patience and grace.
I stand in awe of you. And I thank you.

Time spent working in your garden will not be deducted from your life.

—M. S.

Contents

Author's Note ix

Introduction: Welcome, Everyone xi

1. Victory Gardens Then and Now 1

2. Planning Your Glorious Garden 21

3. Gear Up! 47

4. Designing, Planting, Nurturing 71

5. The Good Stuff: Vegetables 89

6. The Good Stuff: Herbs 147

7. The Good Stuff: Edible Flowers 177

Epilogue: A Ribbon Through Time 205

Acknowledgments: In Gratitude 215

Photo Gallery 217

Appendix: Resources for Gardeners 233

Index 247

About the Author 255

Author's Note

Throughout this book, you will find many places where I refer to specific varieties of plants by name. Sometimes I'm using a particular plant editorially, to illustrate a point. More often, especially in chapters 5, 6, and 7, naming a specific variety represents my recommendation to you.

Occasionally I also include the name of a particular source for those items. When I do this, it's because, as far as I now know, they are not available elsewhere. That could well change in the future, but as of now, that named source is exclusive. If no source is mentioned, that means the items are available from multiple sources. I receive no material benefit from mentioning these companies other than the very real pleasure of introducing you to some wonderful businesses and their products.

In the photo gallery at the back of the book, you will find color photos of many of the recommended plants. Underneath many of those photos is the name of a seed house that provided the photo. In all likelihood, they will still be carrying that product by the time you're reading this book; if not, they can surely suggest a close substitute. And a few minutes with your favorite search engine will reveal other shopping possibilities.

INTRODUCTION

Welcome, Everyone

This is a book about growing good things to eat—vegetables, herbs, and edible flowers—and doing it completely in containers. It's the way I've been gardening for the past twenty-plus years, ever since I moved into a lovely old apartment building with no garden space at all, just a concrete patio about the size of a bandana.

Every year I learned a little more about this type of gardening. Mostly through trial and error, I figured out what works and what doesn't and what isn't worth the trouble. Gradually, too, I began to realize that container gardening is a very solid answer to many different circumstances that people might find themselves in, not just the no-yard folks like me. Plant breeders were noticing the same thing and began developing many wonderful new varieties especially for containers. Everything seemed to be humming along nicely. Then, in 2020 something remarkable happened.

I'm sure you remember 2020. That's the year a mysterious, terrifying virus turned our world upside down and kept many of us essentially trapped at home, afraid to go about our usual lives. Including shopping for groceries.

Then suddenly, the whole world decided to plant a vegetable garden, in whatever space they could find. I found myself thinking about the Victory Gardens of World War II, when people all over the country planted a few vegetables to feed their families. *Were we witnessing a modern-day version?*

So I decided to look into this a little, and gradually I realized I was seeing a remarkable drama repeat itself with uncanny echoes. That drama, if it had a name, might be something like "national trauma plus gardens." We saw it first in World War I, then World War II, and again with the coronavirus pandemic of 2020. You'll find the full story in chapter 1. There you'll also begin to meet some wonderful people whose memories of the Victory Gardens that their families tended during World War II make those days real again and remind us, lest we forget, of the healing power of gardens—evergreen and timeless. I was deeply moved by their stories and feel blessed that they allowed me to share them with you.

In the chapters that follow, you'll learn how to evaluate your space and focus in on the plants that will do well there (and skip the ones that won't), ideas for making the most of your limited space, what tools and equipment you need and what you can do without, solid techniques for planting and growing your favorites, and a lot more. Then the good stuff: plant-by-plant information on specific vegetables, herbs, and edible flowers that I believe will do well in your container garden, even if—especially if—this is your first.

I Wrote This Book for You If . . .

You yearn to grow your own fresh vegetables but . . .

- You don't have a real garden space. There are lots of very reasonable reasons. Maybe you live in an apartment, a townhome, or a condominium. Or on a houseboat. Or you have a perfectly fine house but there is just no gardenable yard space.
- You do have an actual "in the dirt" garden area, but it's too shady for vegetables.
- You aren't as limber as you once were and don't like bending over.
- You live with mobility limitations that make it impossible to get down on the ground.
- You probably *could* get down on the ground, but you'd rather not, thank you very much.

For all those situations—and more—container gardening just might be the perfect answer.

But maybe you don't know how to get started because . . .

- You've always lived in apartments; you've never gardened in your whole life.
- You're a very experienced gardener, but this container thing is brand-new to you.
- You tried it years ago and it was a disaster. It would be like starting all over.
- You did your first container garden as a 2020 Victory Gardener, but the results were disappointing. You suspect you didn't start with good coaching.
- Someone in your family actually had a World War II Victory Garden, but that was many years ago and they're not around to teach you.

Help for Beginners

Now, if you look back over those lists, you may notice something important: no matter their circumstances, all those folks can be considered beginners, in the sense that they are ***new to gardening in containers.*** Even those who've been gardening for years in traditional gardens will immediately recognize that container gardening is quite a different kettle of kale. So when planning this book I made a conscious pledge: my goal here is to help beginners get on the right path from the start.

That perspective guided many decisions as I was writing this book. For one thing, it means that I had to make tough choices about what plants to include. Yes, you can grow practically *anything* in a container garden, but not everything is equally practical or equally satisfying. Container gardeners have limited gardening space; that is a given. Success means finding ways to use that limited space to the fullest. One way to do that is to avoid plants that produce a small amount of foodstuffs in relation to the space that the plant needs: a question of ratio, which I address often because it is such a handy guideline. If you're wondering why a particular plant is not included, that could be the reason.

Or it could be that I decided certain items would be too challenging for beginners. The last thing I want is for new gardeners to become frustrated, and the best way I know to avoid that is to focus on those plants I believe will give them lots of healthy food and great pleasure with a minimum of heartache. I want you to know that none of these decisions was easy.

Maintaining that success-for-beginners perspective also means that when I describe how to do some specific gardening task or technique, I will focus on the one method that I believe beginners will find easiest to master. If you already have a favorite way of doing that task, a different way, by all means keep at it.

It also means that sometimes experienced gardeners will find themselves reading something they already know. I do understand that can be annoying. But if I did *not* take the time to explain that thing, whatever it is, beginning gardeners would be lost, and I consider that much the greater sin. I hope you longtime gardeners will understand and be patient.

To every single one of you—absolute beginner or longtime gardener and everyone in between—I send my greetings and my gratitude. Thanks for taking this journey with me. I hope you'll find useful information here and maybe new inspiration. And if you feel so inclined, I'd love to hear about your experiences.

A Word in Closing

Gardening is hope made real. Planting a seed comes with a promise from the universe that something tangible, something real, will follow from that simple act. And if it's the seed of a food plant, there is a further promise that what will follow is actual nourishment. You, in turn, promise to nurture the young plant into maturity and then selflessly share its bounty.

So consider this: even if it was hard to tackle something new, and even if you sometimes wonder whether it really made a difference in the long run, you and that tiny seed brought life itself from the soil. Don't lose sight of that.

Maggie Stuckey
Portland, Oregon
May 2022

CHAPTER 1

Victory Gardens Then and Now

This is a book about growing a vegetable garden in containers. I started thinking about it in 2020, when—in response to that terrifying virus that kept us afraid to venture outside, even for such essentials as grocery shopping—so many people decided to grow their own vegetables. Many of them had neither gardening experience nor a suitable garden space, and so they just did the best they could.

It's very much like what happened during World War II, when Americans planted a few vegetables in whatever little patch of ground they could find and called it a Victory Garden. Watching the astonishing surge of gardening activity almost eighty years later, I started to wonder: Are we seeing a modern-day version of those wartime Victory Gardens? I'm sort of a history junkie, so I decided to see what I could find out.

The first thing I learned is that I was wrong. I always thought of Victory Gardens as a feature of World War II, but they actually began with World War I, although they had a slightly different name then. Even more surprising was how eerily the tragedies mirrored each other through the decades: World War I with its gardens and its influenza pandemic, World War II with its gardens and its devastating loss of life, 2020's gardens in response to the coronavirus pandemic. Only when we lay them out in sequence do the parallels become so clear, so astonishing.

Then . . .

The year is 1917. War is raging across Europe. Young men are fighting and dying in heartbreaking numbers. Commentators are calling it "the war to end all wars," but later, in retrospect, we will come to know it as the First World War.

Most of Europe is still an agrarian economy. Today's soldiers were yesterday's farmers, and today's battlefields were their croplands, now destroyed. Widespread food shortages are common; soon, Europe may face the very real possibility of sustained hunger, even famine. The United States is not yet an active combatant, but one leading citizen is becoming increasingly alarmed about the food crisis. And he's in a position to do something about it.

Charles Lathrop Pack, timber baron, is the son and grandson of legendary timbermen and continues to run the family business. Decades of careful timberland management have given him an intuitive feel for working with natural resources, and three generations of success have made him one of the five wealthiest people in America. He is haunted by the image of people going hungry because their homelands are trapped in war, and he has used his two strengths (resource management and political influence) to activate an idea: What if Americans were to plant food crops in patches of unused land and contribute some of the produce to Allied armies and civilians? With a patriot's passion and a businessman's leadership skill, he has convinced powerful people in government to support his idea, and in March of this year he announces the creation of the National War Garden Commission.

He serves as the commission president and has recruited people of influence to join him: top scientists such as Luther Burbank, several university presidents, leaders of prominent civic organizations, and high-level representatives of most of the country's primary industries. In the aggregate, they have considerable leverage in Washington.

Much of their work involves creating a wide-ranging public information campaign with just one goal: to encourage people to plant food crops in any little piece of unused land to help provide food for America's allies fighting in Europe. The message is carried in cartoons, short stories, and patriotic essays distributed to newspapers across the country; free handbooks of garden tips; nutrition and food-preservation guides; and inspirational posters created by prominent artists of the day. At the same time, the commission is working hard behind the scenes to make sure the new gardeners get support from businesses, civic organizations, and government agencies.

Then, on April 6, one month after the formal start of the War Garden Commission, the United States declares war on Germany, and the first contingent of American troops ships out. Those good-hearted folks who earlier created tiny pocket gardens to help strangers in faraway Europe now have a new urgency: their efforts might actually help feed American soldiers, maybe even their own sons. And they themselves are proud to be called "soldiers of the soil."

Did it succeed? We shall see.

The year is 1918. The soldiers of the soil work hard. By the time the war ends later this year, they will have created 5,285,000 gardens and delivered produce worth $520 million (more than $9 billion in 2020 dollars).

Meanwhile, the fighting soldiers have suffered unimaginable casualties. Record-keeping is imprecise, and statistical guidelines vary from country to country, so the best we can do is rely on estimates and speak in round numbers. Even so, the totals are staggering. Of a total of some 65 million armed forces, almost 37.5 million suffered casualties—8.5 million killed, 21 million wounded, nearly 8 million prisoners of war or missing. To say that another way, 65 million men went to war, and more than half of them were injured or killed. There is more. Unknown, uncountable, are those broken by what puzzled doctors call *shell shock* and from which many will never fully recover.

On November 11, a day that will later be celebrated as Armistice Day, Germany formally surrenders.

So the war is over, but not the dying. A terrifying new enemy—invisible, silent, deadly—has replaced the machine guns and the mortars. A ferocious influenza pandemic has been roaring through every town and nation, ruthlessly taking the lives of the very young and the very old and, most surprisingly, young, healthy adults. No one knows exactly what it is or where it came from. No one knows how to keep people from getting it or treat them when they do. Vaccines are still decades in the future.

The first wave, back in the spring of this year, had been comparatively mild. Many in the US, including some in the medical field, conclude this is merely the usual flu, nothing to worry about. Tragically, many will continue to hold on to that flawed belief through the summer, even as the pandemic rages on.

Seemingly overnight, this terrible disease has taken over people's lives. Doctors struggle to control a dreadful illness they have never seen and for which there is no treatment. They are particularly astonished at the speed with which people are stricken (sometimes mere hours after exposure) and die (two or three days). To those still skeptical, they tell the story—the *true* story—of four women who played bridge one evening; the following day, three died from influenza. Hospitals, already short-staffed with so many medical personnel serving in the military or themselves stricken with the disease, are quickly overwhelmed. Rows of hospital beds are set up in private homes; storage sheds are turned into makeshift morgues. Everywhere, one medical journal reports, exhausted practitioners are very near the breaking point.

Fear of the unknown is a powerful thing, and many people turn their anxiety into suspicion. Soon, news reports are spreading the rumor that the Germans deliberately created this epidemic. These reports are false, but many people eagerly embrace and repeat them; there is comfort in having someone to blame.

Meanwhile, medical and public health officials continue doing their best. They call for quarantines, and encourage people to wear masks, to refrain from shaking hands, and to strenuously limit their contact with others. But there is no broad-based coordinated government response. Local governments, left to their own devices, respond in many different ways. Some take quick action, shutting down businesses and closing schools and churches. San Francisco promises to fine anyone not wearing a mask in public. St. Louis has banned all public gatherings. Other officials, feeling pressure to maintain a patriotic spirit during wartime, downplay the seriousness of events. What happens in Philadelphia is especially chilling.

The city's public health director, insisting this is just a normal flu, has refused to cancel the upcoming Liberty Loan parade. It helps promote the sale of war bonds, he explains, and besides, he doesn't want to cause a panic. So, on September 28, an estimated 200,000 people crowd together on a downtown street to cheer a two-mile parade of colorful floats and marching bands. Three days later, every bed in the city's thirty-one hospitals is filled with patients suffocating as their lungs fill with mucus and blood. One week after the parade, 2,600 people have died; the next week, 4,500. In a matter of weeks, this disease will take the lives of more than 12,000 citizens of Philadelphia.

The influenza storm will rank as the worst pandemic of the twentieth century. Around the world, many millions will die (estimates range from 20 million to 100 million), including some 675,000 Americans. As a gruesome comparison, the death toll of the war itself is 16 million.

The story of this epidemic is intimately interwoven with the story of the war. Some historians theorize that America's soldiers, unknowingly infected while in Europe, brought the terror home with them. It is also quite likely that the reverse is true—that they took it with them to Europe, where it infected yet more people in a deadly vicious cycle. Certainly the intense crowding at US military bases and training camps, followed by days of very close quarters aboard ship, was a perfect breeding ground.

Even the end of the war doesn't mean the end of the pandemic; people joyfully rip off their masks and celebrate the November armistice with parades and large parties—and no precautions. People continue to die.

The year is 1919. This year brings two milestones that are significant to our story. The first is that by midyear the influenza pandemic finally burns itself out, as all those who had been infected either perished or recovered, which gave the survivors immunity. The second is that the notion of Victory Gardens, by that name, is born—by accident.

The war is over, but Charles Pack is still hard at work. He knows that the armistice agreement does not magically solve the worldwide food crisis, so he starts a new campaign promoting what he calls *Victory Gardens*. He means "gardens in a time of victory," the next logical step after the end of hostilities. He even writes a fanciful description for this evolution: "The War Garden was the chrysalis. The Victory Garden is the butterfly."

He has no way of knowing he has created a name for another wartime gardening movement still twenty years in the future, in another war to end all wars.

Also in 1919, his history of the war garden experience, titled *The War Garden Victorious,* is published. In it he deeply acknowledges all the organizations, industries, civic groups, governmental agencies, and individual citizens who made it work. Many industries were given their own chapter.

For example, chapter 7, "How the Railroads Helped," recounts the responses of the nation's rail companies, who agreed to allow their employees to plant crops on the rights-of-way and often provided the necessary supplies. As one example, Mr. Pack mentions the 2,100 pounds of seed potatoes given out by the Buffalo, Rochester, and Pittsburgh Railway. When those were all planted, he says, the men bought more on their own.

Hold on a minute. Railroads? Potatoes? I have a wonderful story for you. A story about railroads and potatoes and compassion.

While I was immersed in writing this book, I made a short visit to two very dear friends, Jim and Marian Lee. They live in a sweet little house on the Long Beach Peninsula, a minuscule pinkie-finger of land at the southwestern-most corner of Washington state, where the Columbia River shakes hands with the Pacific Ocean. I adore them both, but this is Marian's story, so she takes the stage here. She had recently turned ninety-eight. This is the story she told me.

Photo courtesy of Diana Thompson

MARIAN LEE

From the time she was four until she left to join the navy at twenty-one, Marian lived with her aunt and uncle in their very Italian neighborhood in Renton, Washington. Uncle Constante (called "Con") and Aunt Mary (known as "Mamie") had immigrated from a small village in Italy where everyone tended a garden of vegetables, so in their new home in America they just naturally continued the tradition. Marian thinks they probably didn't call it a Victory Garden, but that's exactly what it was, because all up and down the block, neighbors shared the bounty of their gardens with one another and with families not so fortunate. By the time Marian came to live with them in 1926, the war was over but not the garden. And not the spirit of community, of looking out for others.

But this is a story about potatoes. Even as a youngster, Marian recognized Uncle Con's pride in his job with the Union Pacific railroad, and so she understood what a profound tragedy it was when a traumatic injury at the railyards forced him into retirement. But Uncle Con loved those trains and loved his coworkers, and even after he was no longer actually on the payroll, he spent many afternoons walking the tracks, checking for any trouble spots. Marian and her cousins often tagged along, captivated by the stories he told them as they walked, and watching as he stopped here and there along the railbed to scuff up a small hole and tuck in a few of the seed potatoes he always seemed to have in his pocket.

Later, Marian says, when they went back over the same areas to harvest the potatoes, they often found that others had been there first. But Uncle Con didn't seem angry; he just smiled and said something like, "Well, guess someone needed them more than us." It was many years before Marian realized that was his plan all along—to do what he could to see that the men riding the rails in those rough years had a little something to eat.

Marian's own story continues into World War II. Like most Americans, she had been following the war news carefully, and her mind was made up. So on September 4, 1943, her twenty-first birthday—and the first day she was eligible—she marched into the recruiting office and signed up for the WAVES, the all-volunteer women's branch of the navy. She served in uniform through most of World War II.

Meanwhile, back at home, people were beginning to plant what by then were being called Victory Gardens.

The year is 1941. War has been raging across Europe for more than two years, ever since Britain and France declared war on Germany in 1939. Even before that, European leaders had watched with increasing concern the relentless march of Hitler's armies through Europe. When the Nazi troops invaded Poland on September 1, 1939, despite earlier promises of safety, Britain and France immediately declared war. Later, historians will generally agree that the invasion of Poland marks the start of World War II.

Almost exactly one year later, in a move that will become critical for our story, Germany and its ally Italy formed an alliance with Japan, which had its own expansionist plans in the East. Signing the Tripartite Pact on September 27, 1940, the three nations agreed to offer assistance if any one of them came under attack by a country not already involved in the war. It was a deliberate taunt to the still-neutral United States, daring it to join the Allies in war.

Officially, the United States continued to remain neutral for another fourteen months, although many negotiations were taking place behind the scenes. In the last month of this year, 1941, everything changes.

In the early morning hours of December 7, a Sunday, Japanese fighter planes begin a fierce surprise assault on the US naval base at Pearl Harbor, Hawaii. Two hours later, 18 warships and 164 aircraft have been destroyed and more than 2,400 US troops killed.

The next day, President Roosevelt announces that following this "day of infamy" the United States has declared war with Japan. And because of the Tripartite Pact signed a year ago, this means that the US is now effectively at war with Japan's allies. Three days later, the US officially declares war on Germany and Italy, thus becoming a full combatant in what will later come to be known as the Second World War. Things begin to move quickly.

The year is 1942 (and beyond). For the next four years, American families struggle through life in wartime. Those with loved ones in military service follow the news carefully. Foreign places whose names no one had previously heard of become part of the national conversation: the Ardennes, Iwo Jima, Normandy, Okinawa, Leningrad, Guadalcanal, Kursk, Anzio. Everyone knows where they were on December 7, 1941. Everyone knows someone whose husband, brother, son, father has been killed. Everyone, in their own way, prays for peace.

In the meantime, there is much to be done at home. Women are working in manufacturing plants converted to military production. Schoolchildren are saving their allowances to buy war bonds. Teenagers are organizing neighborhood drives to collect paper, rags, and string for recycling, which will free up manufacturing capacity for military needs. Many adults are signing up to volunteer as air raid wardens, Red Cross workers, and more.

And nearly everyone has some kind of vegetable garden, even if it is nothing more than a small patch stolen from the lawn. They know home gardens reduce the need for trucking

My dad, George Fech, and my uncle, Norman Duddridge, served our country in the US Army in World War II. While they were overseas, all my relatives grew vegetables and flowers in backyard gardens. Our home was in Aurora, Illinois. My grandparents lived nearby in Lombard, and Uncle Norman and Aunt Irene lived in Villa Park. All these little towns are suburbs of Chicago, and there was lots of visiting back and forth among the families; today it is one big urban area.

Many years after the war was over and my father had returned home safely, I asked him about his experiences. He politely declined. He just didn't want to talk about it. I've talked with other veterans who felt the same way. War is hell. So I turned to my mother, Viola Fech. She was trained as an army nurse and was willing to go overseas and serve. She told me she imagined that the setting would have been much like the TV show *M*A*S*H*. The war was winding down during the last year of her training, and she was not called into service. What she did instead was what so many families did—wait, hope, and pray. And try to keep busy.

When I asked my mom if she would tell me about the wartime days from her perspective, she mostly talked about the family gardens. She spoke with such passion, such natural eloquence, that I remember everything practically word for word.

"It always seemed to me that those Victory Gardens had three purposes. Obviously, they provided fresh produce when supplies were scarce. In order to support the troops, we made many sacrifices. It seemed like a lot of most everything went to the war effort—food, clothes, fuel, and lots of other important everyday supplies. We suffered a little bit as a result, but nothing like the troops did. They were the ones seeing the bombs going off and fighting to protect our freedoms at home. We really didn't mind making the sacrifice, because we all shared in it for the good of the country.

"Also, those gardens lifted our spirits. Seeing the green vegetable plants and the pretty flowers, iris and lilacs and columbine, just made you feel better—especially after hearing Paul Harvey on the radio with war reports with words like, 'Sad day today; thirty troops were killed when . . .'

"And in a very real sense they gave us something to do to help us keep our minds on things other than our husbands and sons being in harm's way. You just kept planting, picking, and pruning, and it took your mind off what might be going on in Europe and Africa."

For me, this was such an important piece of history that I felt compelled to share it with my daughters. One of them, Natalie, was so moved that she interviewed my mom for a 4-H project she was working on called "Garden History Interview." They were to interview an older person and create a display of their answers along with a photo for the county fair exhibit. I'm happy to say that she won a purple ribbon for her effort but also learned to appreciate the sacrifices made by her family members.

The family love of gardening runs deep through generations and continues even now that Mom and Dad have passed. In their last few years, they loved to drive from their home in Lincoln, Nebraska, to visit us in Omaha. They would sit on our deck and watch the action in the backyard—the birds, the trees and shrubs gently waving in the wind, even the rabbits eating our flowers. Just a very calming experience for them. Now my wife and I do the same thing.

John Fech
Omaha, Nebraska

food long distances, so resources like fuel and tires can be redirected to the war effort and large agricultural operations can be refocused toward military provisions.

This is the essence of what are now being called Victory Gardens. Growing food for your own family or to share with neighbors becomes a powerful act of patriotism. It therefore seems surprising to later observers that the federal government is not wholly behind this grassroots effort.

Mrs. Eleanor Roosevelt, America's first lady, remembers poignantly the war gardens of World War I. Now Mrs. Roosevelt hopes to plant a vegetable garden on the White House lawn as a symbol of support for the sacrifices so many American families are making. The recently appointed secretary of agriculture, Claude Wickard, thinks it a terrible idea and is not shy about saying so.

Mr. Wickard has come to Washington from the vast farmlands of Indiana. He is certain that the war gardens of World War I had been a major mistake: folks who knew nothing about gardening planted the wrong things in the wrong places and in the process wasted seeds and fertilizer that could have been much better used by farmers who knew what they were doing. He continues his opposition through all of 1942. When the mayor of New York asks him to develop a Victory Garden program for big cities, Wickard refuses. "It is the Department's position," he writes, "that . . . we should not give much encouragement to growing vegetables in the cities."

In the early years of World War II, we were living in Kalona, Iowa. It was a wonderful place to grow up. Kalona was surrounded by Old Order Amish, the best gardeners, and every family for miles around had a Victory Garden. It was a badge of patriotism.

My mother took her patriotism seriously. She planted a huge, 100-foot-by-100-foot Victory Garden with potatoes, beans, tomatoes, corn, peas, strawberries, and the sweetest carrots. My sister and I were the gatherers, but we were only allowed to walk in the garden, not run. If we ran, we might fall, which meant dropping and spoiling whatever harvest we were holding. Losing food was serious, so we knew that if we were found running, it was a definite case for punishment. Our little brother liked to taunt us by running up and down on the edge of the garden. One day, we just couldn't take it anymore.

Next to the garden was a clothesline, and that day my mother had laundry hanging on one half of the clothesline, and on the other half rugs were being aired out; the rug beater was on the ground. Our brother was darting in and out between the clothes and under the rugs, and we grabbed the rug beater and chased the little monster with it! We ended up running right into trouble—our mother—and spilled all the vegetables we had gathered. I was the oldest, so I was punished.

Having a Victory Garden in Iowa was a duty and gave us a sense of pride; I don't recall a single home without one. Our mothers would talk over the fences, mostly about their garden successes. Seeds were shared and bragging abounded. Today I remember those gardens as a place of patriotic pride primarily tended by women doing their part on the home front.

Nancy Draper
Portland, Oregon

But the families of America do not need his encouragement. They plant their gardens anyway, in big cities, small towns, and rural communities. They dig up lawns, convert their driveways to garden beds, put large, plant-filled containers on urban rooftops, and build window boxes for apartments. They lobby local governments to convert public parks and playgrounds to public community gardens, and convince hometown business owners to give up half their parking lots to employee garden spaces. Sales of seeds this year jump 300 percent from 1941.

By 1943, Mr. Wickard has come around. His personal feelings are not made public, but worldwide events have blunted his opposition, and in early spring of 1943 it is announced

For much of my childhood, my father worked as an attorney at a large chemical refinery—International Mineral and Chemical Corporation in East Point, Georgia. During the years of World War II, he directed that a very large Victory Garden be established on the grounds outside the office. The Georgia soil was rich, sandy loam, and that garden produced plenty of tomatoes, beans, corn, and potatoes, plus watermelons in the summertime.

All the produce from that garden was shared among the people working in the office, and I think some of them also helped maintain the garden, along with the company's janitor. Each employee regularly took home a portion of that week's harvest, and in most homes a canning marathon followed.

I can picture kitchens full of women canning the garden produce for wintertime, working in blazing hot weather with huge kettles steaming. It was brutal work, but we all did it. We knew we needed to save every bit of food that we grew. Thinking back, those Victory Gardens stand out in my mind as a time when everyone worked together to produce food for our tables.

Suzanne Briley
Hobe Sound, Florida

that he has "taken personal charge of a campaign" to promote home vegetable gardens, with a goal of eighteen million Victory Gardens by the end of the year.

Once again, American families go ahead on their own. By the end of the year, they are working in some twenty million gardens on more than twenty million acres across the country. Next year, they will do even more: Victory Gardens will provide fifteen billion pounds of vegetables, an astonishing 40 percent of the country's total consumption of fresh produce.

And Mrs. Roosevelt gets her White House garden.

The year is 1945. American troops are fighting a world war on two fronts, half a world apart. At home, families go about their daily lives, quietly accommodating all the changes

I was a small child during the war, but I remember our Victory Garden vividly. We had moved to Longview [Washington] and bought a house with fruit trees on a one-acre lot. We had tomatoes, carrots, peas, beets, corn, and beans, both green beans and "shelly" beans. Also raspberries and strawberries. We canned everything we did not eat immediately. My brother and I worked in the garden, with our grandmother supervising. My brother hated being dirty, hated bugs, and hated gardening in general. I loved it.

One thing about that time that I will never forget is one special day. Two days, actually. On V-E Day, and again on V-J Day, my family gathered in the living room to listen to the radio news about the end of the war. As "The Star-Spangled Banner" played, we all stood with our hands on our hearts.

Margaret Staudenraus
Long Beach, Washington

that wartime brings: Women work factory shifts at the jobs once held by men now serving overseas. Gasoline is rationed, so use of the family car is sharply reduced. A good thing, too, since no new cars have been manufactured since 1943, when those factories retooled to produce military items. Facing shortages of household staples, people develop ingenious substitutions. They take pride in making clever use of leftovers. They have even finally figured out how to use those darn ration books.

And they continue with their gardens. By the time the war ends this year, Victory Gardeners will have grown more than one billion tons of food. Throughout, in their own various words, those tending the gardens have expressed the same sentiment: *To me this garden represents life. It means that my husband/son/father/brother is coming home safe.*

On May 8, Germany surrenders. On September 2, Japan surrenders. The second war to end all wars is over. Worldwide, some 15 million have been killed in battle, an additional 25 million wounded. American families have lost 416,800 husbands and sons and brothers.

. . . And Now

Is it true, as the pundits like to remind us, that there is nothing new under the sun, that history will always repeat itself and we ignore its lessons at our peril? Maybe. But most days I'd rather leave lofty questions like that to the philosophers and the poets, and instead think about the small, everyday acts of kindness and grace that illuminate our common humanity. They always strike me as both new and, at the same time, timeless.

Seen from that perspective, what happened to us in the second decade of the twenty-first century is both new and not-new, surprising and eerily familiar. A century after we endured, and mostly survived, a devastating pandemic, we were hit with another. And once again, we found solace in the garden.

The year is 2019. Precisely one hundred years after the world's worst recorded pandemic ended, a new one arrives. The coincidence of timing is almost unbelievable; if this were a movie, you would roll your eyes. Even more amazing is how closely the new events, as they unfold, will echo the old.

GROW VITAMINS AT
YOUR KITCHEN DOOR

The origins of the disease—exactly where and when and how—are not completely clear. Much of that story is overlaid with rumor and suspicion, further complicated by the intricacies of diplomatic relations between less than friendly nations. We do know a few things.

We know that the earliest reported cases, in November, are in China. We know that those first incidents are thought to be a new and particularly severe type of pneumonia. We know that in December a man in Paris, who had not traveled to Asia or had contact with anyone who did, contracts the same mysterious illness. We know that, as infection rates climb, late in December the World Health Organization steps in, and on December 31 announces an "urgent, serious" investigation into what is still considered an outbreak of pneumonia.

That investigation ultimately will bring some answers. Medically, it is a coronavirus, the general name for a large group of viruses that infect the respiratory system, some quite serious, some not. This one shows some characteristics heretofore unknown, and so it is termed the "2019 novel coronavirus," *novel* meaning *new*. The disease that it causes is given the name COVID-19, designating the year of its discovery.

Next year, the devastation it causes will reach unimaginable levels.

The year is 2020. Once again, a deadly disease no one has ever seen before has set the world afire. In a very real sense, we are once again at war, fighting for our survival against an unknown enemy. It's a new kind of war, but at the same time strangely familiar, for it brings its own food crisis. And once again Americans respond with the age-old, life-affirming act of planting a garden.

In the early days of this year, information about this mysterious disease is murky; it seems to trickle in and is frequently contradictory. Epidemiologists struggle to get a handle on the full scope of the situation, while some political leaders insist this is just a normal flu, nothing to worry about. Tragically, many will continue to hold on to that flawed belief through the summer, disregarding safety guidelines even as the epidemic explodes.

In March, the World Health Organization declares this disease a pandemic. By spring, it has taken over our lives.

Doctors struggle to control a terrifying illness they have never seen and for which there is no vaccine. The infection spreads quickly, ruthlessly. Hospitals are quickly overwhelmed.

Patient beds fill the corridors, doctors work long shifts for days on end, and nurses wear protective gowns made of large garbage bags. In the parking lots, tightly packed rows of refrigerated trucks normally used to transport perishable goods are turned into makeshift morgues. Exhausted practitioners are very near the breaking point; before the year is over, many will succumb to the disease themselves.

Public health officials are doing their best. They call for quarantines and encourage people to wear masks, to refrain from shaking hands, and to strenuously limit their contact with others. But there is no broad-based coordinated government response. State and local officials are left to their own devices, competing with one another for resources and struggling to find safety guidelines that their communities will embrace. Some take quick action, shutting down businesses, closing schools and churches, and banning large gatherings. Others continue to downplay the seriousness of events. They wanted to avoid panic, they will later say.

But events overtake obfuscation. As the summer months unfold, more and more people become gravely ill, more communal facilities like nursing homes are completely locked down, more refrigerated trucks are needed. Medical experts race to find treatments for a disease that is still largely an enigma. Confusion, uncertainty, anger, denial, disbelief, fear, panic—people in every part of the country cycle through vivid, paralyzing emotions every day.

Fear of the unknown is a powerful thing, and many people turn their anxiety into suspicion. Certain news outlets and social media groups begin reporting the rumor that the Chinese government created the virus and turned it loose on the world. In some places, suspicion begets violence, and Americans of Asian heritage are assaulted on the street.

Others—many others—are choosing a different way.

From almost the beginning, one commonsense safety recommendation has been repeated practically every day: since we don't know yet how this disease is spread, the best strategy is to limit contact with others as much as humanly possible. In practice, this means staying home. And that means, among many other changes, avoiding the grocery store.

Then, almost like magic, millions of people all across the country have the same brilliant idea at the very same time: *OK then, I'll just grow my own groceries.* The result is a surge of interest in vegetable gardening the likes of which few people have ever seen.

Frenzy might be a better word. Practically overnight, mail-order seed companies are flooded with more orders than they can handle. Garden centers and nurseries have nothing but empty racks and shelves—and exhausted staff trying to explain. Garden websites crash; so do the phone lines at county extension offices, a popular source of information in normal times. And all of them are doing their best to juggle questions from new gardeners desperately looking for help.

There are plenty of them. A year-end survey from the leading industry organization finds that this year 18.3 million people started gardening for the first time. There may actually be more; one longtime plantsman, looking at his year-end numbers, thinks the total is more like twenty million. That survey told us a great deal about these new gardeners. Most of them are primarily interested in food gardens, and they are mostly young (under thirty-five), mostly male, mostly parents of young children, mostly working full time, and mostly urban. In other words, considering all those factors: short on time and short on space. It's no wonder, then, that more than one in three of them grow their vegetable gardens in containers.

Containers on the balcony, on the patio, on the deck, on the front steps, even on the fire escape (don't tell anyone)—every tiny space available to these young urbanites is now being used to grow healthy food for the family table. Very much like, in spirit if not in physical details, what their great-grandparents did during the war to end all wars: with gratitude and hope, planting seeds in every bit of space they could find.

There is one big difference. Those earlier wartime gardeners could point to the very day the conflict ended. Gardeners in 2020 can only watch and worry as the enemy they fight bulldozes onward. By year's end, more than twenty million Americans will become ill with COVID-19, and 350,000 will die.

Next year will be worse. Tragically, terrifyingly worse.

And in the year after that, we will hit a heartbreaking milestone: on May 17, 2022, the total US death toll from COVID-19 will pass the one million mark.

CHAPTER 2

Planning Your Glorious Garden

You're exploring this book (thank you) because you're interested in learning about tending a vegetable garden in containers. You might be a longtime gardener, full of experience and wisdom, but you've never tried container gardening. Or you might be an absolute beginner, eager to get started but with no real sense of how to go about it. Or something in between.

For all of you, I make this promise: you will be amazed at the amount of foodstuffs you can grow in just a few carefully planned containers.

I also promise you this: growing in containers is easier than growing the same plants in the ground. A *lot* easier.

And this: it is also wonderfully fulfilling. If you've never watched a seed nudge its way up through the soil, proudly grow into a recognizable plant and then into a delicious morsel for tonight's dinner, you have a grand experience ahead. Plus, you're going to have some serious bragging rights with your gardenless friends.

In the chapters ahead, you will learn the nuts and bolts of containers, soil, fertilizers, watering, all that stuff. You will then find very detailed information about individual plants and specific varieties of each one: vegetables, herbs, and edible flowers. (Note: I often use the word *vegetables* as a kind of shorthand for all three of these categories together so I don't have to keep saying *vegetables and herbs and edible flowers* over and over.) This plant-by-plant guide is the heart of the book.

But first I want to give you a sense of just how amazing this way of gardening is.

Let's daydream a little. We'll start by imagining creating a basic Victory Garden of family favorites—nothing fancy, just a good selection of standards so you can put fresh veggies on the table without having to rely completely on the grocery store.

Let's say you have a sunny space for your new garden: patio, deck, balcony, back porch, front porch, even the front stoop will work. Start with the biggest container you can manage; we'll assume it's round, since most are. Visually separate it into three concentric circles. In the spring, put several small lettuce plants in the outermost circle, around the edges, maybe alternating redleaf and greenleaf types. In between the lettuces put several viola plants. In the middle circle, between the lettuce and the center, plant seeds of radishes and small carrots; leave the center bare for now. The radishes will sprout first and will grow fast; the delicate foliage of carrots is very distinctive, so if the two seeds are mixed together (which is actually smart, for reasons you will learn later), you can easily tell them apart.

When the weather turns seriously warm, add one determinate tomato plant in the center. (In the tomato section of chapter 5 I'll explain *determinate* versus *indeterminate* and why that's important; for now, just take my word for it.) Snuggle one or two basils in next to the tomato. The hot weather will soon be the end of the lettuces and probably the violas too, so you might as well pull them out. In their place, add seeds of bush beans. Keep adding radish seeds here and there in any bare spot. And now you can begin to imagine how wonderful it's going to be when you start enjoying the results. Like this . . .

In the late spring twilight, you step outside and carefully remove several outer leaves of both kinds of lettuce, enough for tonight's salad. Gently pull up a few radishes (aren't they cute?) and give a carrot plant a tug; are any of them ready? Then, your special touch: half a dozen delicate viola blossoms to bedazzle the finished salad.

As summertime settles in, you might be inspired to make a fresh pasta sauce from tomatoes and basil you just harvested this afternoon. Holy ragu, it's good! Another night, fresh green beans roasted on your tabletop grill and garnished with beautiful red radishes sliced thin as silk. Maybe served side by side with your version of caprese salad—thick slices of tomatoes, picked just minutes ago, dead ripe and juicy, sprinkled with tiny basil leaves and dusted with kosher salt. And for the kids, their own bowls of those little round carrots you helped them dig up this afternoon.

You did this. Your own Victory Garden. Your personal victory over stress. And that's just one container. With two or three more, you can expand into other vegetables your family likes and maybe try something new as well.

You have probably noticed that I'm in favor of packing your containers full of different types of plants. I also strongly recommend rotating cool-season and hot-season plants in and out of the same space as the year unfolds and the weather changes (see details in chapter 4); the farmer's name for that is *succession planting.* Both of these approaches are meant to help you get maximum productivity from your container garden, where *by definition* you have very limited space.

However, it is 100 percent OK with me if you'd rather fill an entire container with just one type of plant that you and your family absolutely love. It's your garden, sweetheart. I want this new adventure to bring you joy, not more stress from worrying you're not doing it "right." *Right* is whatever works for you, including new things you learn along the way.

Have I convinced you? Good. Let's get you started.

Planning for Success

Planning a garden, like most everything else in life, is an exercise in compromise: balancing what we want with what is actually doable. It's true for all gardeners, but even more so for container gardeners because we have such limited space to work with and thus little room for error. That's why it's important you take the time to think everything through in the beginning, before you start planting something that simply isn't going to work.

To start, here's an overview of the general process.

- Analyze the physical character of the space you're planning to use for your garden. How much sunlight does it get? Do you have a good source of water? Is the area prone to strong winds?
- What vegetables (shorthand term, remember) are you interested in growing? Make a wish list, and make it as long as you want.
- What growing conditions do those plants need? Sunlight, water, any special attention? This will take some serious research and, yes, I'm going to help.
- How does all that stack up against the traits of your potential garden space? Compare all three variables: your physical space with your wish list of plants and your research into those plants. Do you need to make any changes?
- Make sure you understand the usual weather patterns where you live. Will you have

to deal with extreme heat or cold? Then check to see how many plants on your list grow best in cool seasons and how many need warm weather. If your weather patterns don't fit your wishes, you'll need to adjust your list.

- Do you know which plants on your list could logically be described as *easy care* and which are *high-maintenance*? Does that align with your personal time availability? Study your list again. Adjust as necessary.
- When it comes to new foods, how adventurous are you and others in your household?
- If you've had to drop some things from your list, can you find alternatives? This could be very important.

You get the idea. With some soul-searching and good research, build a list and refine it as many times as it takes. Always leave room to try something new, but be honest. If no one in the family likes okra, they're not going to suddenly love it just because you grew it. Some things will turn out spectacularly, others will disappoint. Keep notes and carry on; it's good information for next year.

That's the big picture. Now we're going to look more deeply at the most critical pieces.

Know Your Space

SUNLIGHT

It's quite possible you never paid much attention to this question, but it's the most important of all. Every plant on earth needs two things for survival: sunlight and water. You can add water, but you can't create sunlight. You have to work with what you have.

So start noting sun patterns, which means paying attention to several factors. Which compass direction does your garden space face? A north-facing area will give you the least amount of sun, then east, then west, then south the most. Look for nearby tall buildings or large trees that cast shade on your garden space. How about the shade factor of the ceiling of your balcony or front porch? Try to reconstruct a full year's pattern; do the best you can to think back to last year, and mentally put yourself in that garden space every season: Was it sunny, partly sunny, or mostly shady?

Why does this matter? Because more than any other single factor, it determines which plants you can successfully grow in your space and which will disappoint you. All plants need sunlight for photosynthesis, but some need more than others. It's especially critical for vegetable gardeners.

As a simple rule of thumb, if a plant makes a flower before it makes the part you eat, you'll need six hours of sunlight a day. And when you think about it, that describes many of our common vegetables, not to mention sun-loving herbs and edible flowers.

If you now realize your garden space is mostly shady, or mixed sun and shade, it's not the end of the world. You can still have a wonderful vegetable garden; you just have to plan more carefully. Concentrate on those vegetables for which we eat the leaves, like lettuce, spinach, chard, and all the other wonderful leafy greens; and the plants for which the parts we eat grow underground, like beets, radishes, carrots, and so forth. And you can also cheat a bit on the rule of thumb, as long as you accept that your results will be less than robust. For example, I know you're going to want some big, beautiful tomatoes, so put in one plant and baby it along; even a few is better than none.

WATER AVAILABILITY

Immutable fact of life: plants grown in containers need more frequent watering than those same plants grown in the ground. Make your peace with that, and plan your garden accordingly.

Where is your source of water? If your garden space is part of your own home, like a patio or deck, you probably have an outdoor faucet somewhere. If you need a longer hose, now's the time. A more substantial solution (besides paying the teenagers next door) is a drip watering system on a timer. These systems have improved significantly in recent years, with lower costs.

If you live in an apartment or condo, and your unit does not have an outdoor water source, you'll have to carry the water out to the containers from inside. Don't fret—plenty of people do it. Just keep it in mind and use common sense: put the plants that need the most water closest to the water source. In the next chapter, you'll find some ideas about ways to make this perpetual problem a bit easier.

WIND

There are few things more frustrating than coming outside and finding your beautiful container, overflowing with gorgeous plants, smashed to smithereens on the concrete floor of your balcony. Wind is the culprit here, but it is actually pretty simple to avoid with some

Victory Gardeners didn't always have the luxury of planning their space; they used whatever was available to them, like a nearby vacant lot—or a cemetery.

I was eight years old when World War II came to a close, just old enough to have vivid memories of my parents' Victory Garden in Omaha, Nebraska. Also, both sets of grandparents lived in Missouri, and all of them had huge gardens too. So did just about everyone we knew in Omaha, and lots of Missouri neighbors. It was just what you did back then.

My father was a produce buyer at Safeway, and my mother worked as a welder on airplanes at Gate City Steel. She was small and could get into areas that others could not. But whenever they were not at work, you could usually find some of us working in the garden, because the vegetables we grew were so important for a family of six.

The garden plot itself was not in our yard but on a vacant lot nearby, a short walk away. One day, one awful day, I was helping my father gather potatoes. He had a gunny-sack and a spade, and I was carrying a bucket. The first potatoes he dug up were hard as rocks; he moved to another spot, and then another, and another—all the same. The potatoes had developed some kind of disease, and every single one of them was inedible. It was a big shock and a really serious thing, as potatoes were a main part of our diet at that time.

We managed, somehow, and my dad never lost his love of gardening. He had a small garden in his backyard until the day he died.

Marti Benner
Portland, Oregon

Today our grandkids might think planting vegetables in a cemetery is something out of the *Addams Family*, but as it turns out, with space scarce for a garden in an urban environment, the abode of the dead was an ideal site for nurturing tasty veggies during World War II.

Our father was too old for the draft, but as a steelworker in the Duquesne, Pennsylvania, mill with two toddlers, he qualified for an apartment in the newly constructed Burns Heights housing project, a barracks-style development originally designed for military housing. The long, narrow buildings were aligned in rows and had the latest in fireproofing technology—asbestos tiles and shingles—to protect against the coal-burning furnace in each apartment. With a coal bin in front and a tiny backyard for rotating laundry lines, there was no room for a Victory Garden.

My parents were raised in a farming village in what is now the Slovak Republic, so they jumped at the chance to grab a garden plot offered by the St. Joseph church, in the cemetery. With the plots came convenient water taps for our hose connections along with free water. The grassy cemetery soil had been leveled, fertilized, watered, and tended for years, so it was a ready and nourishing medium for our packets of seeds. And it was only a mile walk from the housing project.

The best part was the harvest. After our mother's canning campaign in the fall, my sister and I could hitch up our little red wagon and sell the excess to the neighbors. By the war's end, my father had saved enough to build his dream house with a large garden in the backyard.

Fred Fraikor
Golden, Colorado

preplanning. Think back to your assessment steps: Did you remember days when your potential garden area was hit with strong winds? Even one is too many, so plan your strategy now, before you plant anything.

Some ideas:

- Use the largest containers you can manage; when filled, their weight will be preventive all by themselves. For extra safety, before you add soil, put something heavy in the bottom, such as a couple of bricks or fat rocks. (But do *not* block the drainage hole.)
- Position your vulnerable containers as close to a wall as possible; that automatically cuts down on half the wind force.
- Attach a strong cord or wire to the container and then fasten that wire to a solid element of the building, like a balcony rail or roof post. If the container is wood or plastic, drill two holes, thread your wire through and back out, snug it down to itself, and then tie onto the support. If the container is something you can't drill, thread the wire down from the top edge to the bottom, through the drainage hole, then back up along the outside; tie it off to itself at the top, and tie the remaining long end to the support piece. Easiest of all is a container that has some element you can fasten your cord to directly. See drawing on page 30.

Know Your Weather

For gardeners of all stripes, most of the angst about weather centers around cold temperatures: Do your plants have the genetic makeup to survive winter's cold, or will they perish at the first freeze? There are two common ways to describe this cold tolerance, and you should be aware of both.

USDA ZONES

The first, and most common, is the system of *hardiness zones* established (and regularly refined) by the US Department of Agriculture. These zones (currently, there are thirteen) are delineated by the *average coldest temperatures* and are expressed as a number; the lower the number, the lower the winter temperature in that zone. South Florida is zone 10, parts of Alaska are zone 1.

Let's say that again: USDA zones identify only one factor—the average coldest temperatures, the point at which plants are unlikely to survive the winter. For vegetable gardeners, this is largely academic, since most of our plants are annuals and so are not expected to live through the winter anyway.

However, you need to know about the USDA system because it is so well established that much general gardening information for consumers includes that information; you are bound to run across it and wonder what the heck it's all about.

SUNSHINE RULE OF THUMB

If a vegetable plant makes a flower before it makes the part we eat, that plant wants six hours of sunlight a day.

If you want to know your zone, because you're curious or want to impress your new gardening buddies, it's extremely easy to find; just ask your favorite search engine for "USDA zone by zip code." If you receive seed catalogs in the spring, your zone number might be on your mailing label.

Just remember: the zone number is one-dimensional information. Let me show you how limiting it can be. I grew up in South Carolina, with long, hot summers, long days of sunshine, and relatively mild winters. I currently live in western Oregon, which even though it's far north on a map actually also has mild winters. We also have lots of drizzly days with gray, sunless skies and coolish weather that doesn't really warm up until at least June. Many of the vegetables I remember from my grandmother's South Carolina garden would not be successful in Oregon; the growing season is simply not long enough. And guess what—because of their winter temperature patterns, they're both in the same zone.

FROST DATES

More useful to home gardeners, to my way of thinking, are two other numbers: the spring and fall frost dates. That means the last day in the spring and the first day in the fall that, on average, the temperature where you live hits thirty-two degrees. The time in between is your growing season.

Thanks to the internet, it's easy to find the frost dates for your area. Using your favorite search engine, ask for "frost dates by zip code." You might also investigate microclimates of your immediate area; ask a knowledgeable neighbor, garden club or Master Gardener group, or county extension office. In other words, the more local, the better.

Outsmarting Your Environment

Remember I promised, at the start of this chapter, that container gardening is easier than the kind your friends with big backyard gardens do? Well, here's a perfect example. Neither one of you can *control* the environment, but unlike your big-garden friend, you can outsmart it. You can, whenever needed, simply pick your garden up and move it.

Worried about sunlight? Too much or too little? Remember that you can freely move your garden, or parts of it, to follow the sun as its path changes through the seasons. Same idea with weather. A surprise cold snap is predicted for tomorrow? Move that baby up close against the house or even into the garage. Unseasonable heat wave? Move vulnerable containers into the shade, and carefully attend to your watering. And so on.

Three ways to anchor containers against strong winds, left to right: (1) For containers you can't drill, thread the wire or cord down through the drainage hole and back up to the top. (2) Drill two holes near the top, thread the cord through them and back out. (3) Tie cord directly to container where possible (here, loop handles of a grow bag).

Choose the Right Plants

The next big step in getting off on the right foot is choosing the right plants, the ones most likely to be successful in a container garden. Fortunately, because of so many wonderful new varieties on the market, this is light-years easier than it was just a short time ago.

Start with a list of all the things you want to grow. Ask others in the household for ideas, and make it as long as it needs to be. Nobody has veto power, not yet. Be sure to add at least one new thing you think it would be fun to try. Then start narrowing down your list, saying goodbye to suggestions that are simply not practical (your son wants giant pumpkins) and to the things that, on second thought, no one really likes (remember that okra). You'll still have a pretty long list, but it's made up of generalities: carrots, green beans, cherry tomatoes, basil, like that. Your task now is to zero in on the specific varieties of each one that are particularly suited to container growing; if you can't find such a variety for something, that item comes off the list too.

But how to go about finding the best ones when there are so many possibilities in the catalogs and at the garden center? Over the years, I've gradually developed a few guidelines, a sort of mental checklist, that may be helpful at this point.

CONTAINER PLANT CHECKLIST

- Does it have specific varieties that are appropriate for containers, in growth habit and mature size?
- Does it have a good ratio of food produced in relation to the garden space it needs?
- Is it worth the trouble?
- Is it something special you can't easily find at the grocery store?

Now let's look at the individual questions in more detail.

WHAT MAKES A GOOD CONTAINER PLANT

Container appropriate. Here's some very good news: in recent years, the entire horticulture industry has enthusiastically embraced the idea of growing a vegetable garden in containers, and growers have developed many new varieties specifically for us. And every year there are more.

This means that, for just about any vegetable you can think of, there's at least one (and probably more) variety that will do well in your container garden. Your job is to make sure you select *those* varieties and not the standard versions intended for full-size gardens. This is important.

Where to find them? In the catalogs (online or in print) of the seed companies that sell by mail order. There are more of these companies than you may realize; in the appendix at the end of this book, you'll find some of my own favorites, but that is not a complete list by any means.

To be clear: at this point, you are using the catalogs to learn, dream, and start building a wish list. You may well end up ordering seeds from these wonderful folks, and I very much hope you do; but you might also find that exact same item as a baby plant in the nursery later on. This leads you to a complex bit of timing, which I've called the Gardener's Catch-22. See page 39 for more on this.

One thing to keep in mind: horticulture, like any other industry, is subject to the law of supply and demand. Some new things get introduced, don't find a fan base, and quietly fade away. Longtime classics that we assumed will always be with us are suddenly not available, for any one of a number of supply-issue reasons. Or a newer version shows up, eclipsing the old favorite. Or suddenly everyone is talking about a new variety and no seeds are to be found anywhere. Mail-order seed companies will let you know if something you asked for is not available and will likely offer you a very close substitute. Staff at the garden center can usually point you to a good replacement as well. All of which is one more reason to make your wish list thorough—your first choices, then several alternatives for each.

THE MAGIC RATIO

As a container gardener, you have very limited garden space. That's a fact. Some garden vegetables grow on plants that take up a lot of space in relation to the edible things they produce. That's a fact. And that's not good.

To maximize your limited space, look to plants with a good ratio of edible parts to overall size. Best ratio of all: things where we eat the entire plant, like leafy greens. That's the magic ratio.

Favorable food-to-plant ratio. This is one important way you can maximize your limited gardening space. Read the catalog descriptions carefully; pay attention to what they say about mature size, in height and width.

To give you a mental picture of the ratio concept, here are a few examples. And by the way, you won't find the bad-ratio ones anywhere else in this book for the obvious reason that they're wrong for containers.

Bad ratio

- Daylilies. The flowers are edible, but each one lasts on the plant only a day (hence the name), and for containers the plant is something of a space hog.
- Rhubarb. A very handsome plant, but its leaves are the size of turkey platters and besides they are *poisonous*; only the stems are edible.

Extremely bad ratio

- Cabbage, that big green bowling ball from which you make cole slaw. Have you ever seen cabbage growing in the field? It's *way* bigger than the trimmed-out version in the supermarket. A mature plant will take up the entire space of one large container all by itself, will need an entire growing season to reach harvesting size, and from all that you get one cabbage. One.

Good ratio

- Compact cherry tomatoes
- Bush beans or peas
- Beets and radishes
- Any of the leafy greens (because we eat the whole thing)
- Almost all herbs
- In fact, just about everything in the plant chapters later in this book; that's why I chose them!

Worth the trouble? To make best use of your limited garden space, ask yourself if the thing you're considering is really worth the trouble. You can grow those familiar big, long carrots, but it's difficult and takes forever, all for something you can get at Safeway every day of the year. Same with those large russet baking potatoes; too much growing

I'm sorry to tell you this, but Mother Nature is actually in control of your weather. Just ask Clay Nichols.

In 1945 the Nichols family lived in near-poverty conditions at an oil field camp operated by the City Service Oil Company in the woods east of Oklahoma City. With the outbreak of the war, my dad had volunteered for the army, but being a petroleum geologist, he was told to stay home and find oil. The camp consisted of a row of eight company-owned houses and a Victory Garden on common ground shared by all the families. Thanks to ample irrigation water and the hot Oklahoma sun, the corn, peas, potatoes, and tomatoes flourished. I was six years old, but the Victory Garden, together with our scrap-metal collection, was a constant reminder of the war effort. Another was the constant viewing of military aircraft, as the camp was located just southwest of Tinker Field, the large air force base at Midwest City, Oklahoma.

Our Victory Garden met a dramatic and untimely end. On the afternoon of the day that President Franklin D. Roosevelt died, April 12, 1945, my first-grade teacher spotted an approaching storm just as Valley Brook School was ending its day. "Run home, children! A bad storm's coming!" I never ran so fast.

I had not been home five minutes before one of my older brothers came running to our house screaming, "A tornado's coming!" We ran to the community storm cellar just behind our house, a precious few minutes ahead of a mile-wide black cloud touching the ground as it plowed through the oak forest just south of the camp. !

That F-5 tornado damaged our schoolhouse and sent one of my classmate's homes into the nearby creek, then turned just east of our camp, causing significant loss of life. Most of the houses in our camp survived, but Tinker Field was not so fortunate: a number of the B-29 bombers being serviced at the air base were damaged or destroyed.

There was one other casualty of that tornado: our garden. The vegetable plants proved too fragile for the wind and hail. That terminated our Victory Garden permanently, as the war also ended soon after.

Clayton Nichols
Sandpoint, Idaho

time for something so readily available elsewhere. And you already know what I think of cabbage.

Not in the supermarket. OK, you're not going to use your precious container space for things you can easily find in the supermarket. You're going to do the opposite: use your personal garden to grow items you'll *never* find in the supermarket. This is the fun of it all.

I'm talking about things that are just as easy to grow as the "standard" versions (maybe even easier) but will give you that great pleasure of being able to offer something special for dinner. Like tiny Thumbelina carrots, cute as a button and not much bigger. Or pink radishes, with their gorgeous magenta color. Or Chioggia beets, with their beautiful pattern of concentric red and white circles hidden until you slice them. Or mizuna, a beautiful Asian mustard that is delicious raw or cooked, absolutely gorgeous in the garden, and no more difficult to grow than ordinary lettuce.

RESEARCHING IS EASIER THAN EVER

With those general guidelines somewhere in the back of your mind, start zeroing in on good possibilities for your personal list. Of course, you'll want to be sure you're looking at specific varieties that will thrive in your climate; that's where good local information sources (see the following pages) are invaluable. But I hope you'll also think in term of aesthetics. Look for rich colors: yellow tomatoes, purple kale, garnet-toned lettuce, and so on. Don't overlook texture: sleek tomatoes, crinkly oregano, frilly carrot tops. Pay attention to growth habit, which means the overall shape of the plant, and strive for variety: some tall, spiky things; some round, fluffy things; some low-growing things to tumble over the sides.

All of this, and much more besides, is available at the touch of a few keys. Nowadays most companies that sell gardening materials by mail order have online catalogs; some have switched to online exclusively. While many longtime gardeners lament the loss of print catalogs, and sometimes I join them, the fact is that these companies can now provide a depth of information about each and every item that was never possible with printed catalogs. They want you to be satisfied with your choices, so they go to great lengths to provide you with everything you would need to know about growing each plant. It is, without doubt, a treasure trove of knowledge and good coaching.

CHOOSING CONTAINER VARIETIES: TWO RELIABLE SHORTCUTS

1. Zero in on All-America Selections (AAS) winners; any All-America Selection is almost certain to be successful in your garden.
2. Look for the container collections assembled by many mail-order seed companies; they've done the research for you.

Start with these shortcuts. If this is all new to you, and even if it's not, here are two strategies that will help you quickly narrow down the search process into a reasonable starting point. Then from that base you can expand outward in whatever direction strikes your fancy.

All-America Selections is a unique organization and a powerful friend to home gardeners. Under their strict guidelines, each year many new, never-before-sold varieties are trialed for a full season by volunteer horticulture professionals all around the country. At the end of the trials, only those few that show superior performance are designated AAS winners. Starting the following season, seeds of those plants will be available from many online seed sources.

You can find them through the All-America website (see appendix). Just enter your basic interest in the search bar: tomatoes, nasturtiums, whatever; that will show you all the AAS winners for tomatoes, nasturtiums, and so on. (As I write this, it is not yet possible to sort for the "container appropriate" items, but they plan to offer that feature soon.) If you then open up the individual page for a specific tomato or nasturtium, you will find, among other very useful basic info, a tab labeled "buy from." Or approach from the other side: ask your favorite search engine for "tomatoes AAS winners." Or if you already have a favorite seed house, go to their website and ask for "AAS tomatoes." Many seed companies make "AAS winners" one of the main tab choices on their home page; that's a great place to start.

Container collections. Many of the wonderful companies that sell seeds by mail order, ever mindful of helping customers find what they're looking for, have assembled a preselected set of seed packets of items that are especially appropriate for containers. Check in with your favorite seed catalogs, put "container collection" in the search box, and there you are.

Generally there are four or five different seeds in the collection. Then you have two choices: you can order the set as is, or think of it as a well-curated research tool. Here's what I mean by that latter idea. Suppose this is a collection of five items, one of which is a new spinach. But truth to tell, nobody in your household really likes spinach. So open up the individual listings for the other four and read the descriptions carefully. Do they seem appealing, a good fit with your wish list? If so, order those items individually. If one of the four (lettuce, let's say) doesn't seem quite right but overall you like lettuce, widen your search one level and look at all the other lettuces this company offers. And don't blame me if you get hopelessly sidetracked by other lovelies in the process.

Right Plants in the Right Season

There's one more piece to the puzzle, but it's an easy one: matching your plant choices to their best seasonal weather period. The reason it's easy is that there are only two variables: cool-season and warm-season. All the items on your wish list have a genetic preference for one or the other, and that's the time when they thrive. The other times, they will either lose vitality or conk out completely.

Consider lettuce, a cool-season plant. If you plant lettuce in June, you're just asking for heartbreak. Even if you can manage to get it going, at some point (and it seems to happen overnight) the plant will bolt, meaning that it suddenly rockets up into a gigantic, stretched-out version of itself with leaves so bitter they're inedible (see illustration on page 113). There's nothing to do but yank that thing out and promise to pay more attention next year.

I probably shouldn't say that all plants have this limitation, because a few seem to be a bit more tolerant of temperature changes than others and don't show this dramatic reaction to change. Some can straddle seasons successfully. Some specific varieties have been carefully bred for extra performance. "Slow to bolt," for instance, would mean that particular lettuce variety handles hot weather a bit better than others. But as a general rule, plant cool-season plants in cool seasons (which means spring *and* fall) and warm-season plants when the weather warms up (nighttime temperature above fifty degrees is a good threshold).

VEGETABLES	COOL SEASON	WARM SEASON
Asian greens	X	
Beans		X
Beets	X	x
Carrots	X	X
Leafy Greens	X	x
Lettuce and other salad greens	X	
Peas	X	
Radishes	X	x
Tomatoes		X

HERBS	COOL SEASON	WARM SEASON
Basil		X
Parsley	X	

EDIBLE FLOWERS	COOL SEASON	WARM SEASON
Begonias	x	X
Dianthus	X	
Nasturtiums		X
Pansy and Viola	X	

Note: An *X* in both columns means that plant produces in both seasons; a smaller *x* indicates a less robust result. Since most herbs are perennials, for them the season question doesn't apply.

Seeds or Baby Plants?

One issue that bedevils many new gardeners is whether they should start with seeds or baby plants. (And by the way, what do we call those baby plants, anyway? *Seedlings*? *Plantlets*? In some parts of the country they are known as *starts*, which leads to the mildly goofy question, "Do I start with starts?" The most common term is *transplant*, so that's what we'll use hereafter.) The decision hinges on two related factors: timing and availability. Getting the timing right is a question of planning, which is what this chapter is about. The actual how-tos come in the next chapter.

To simplify matters, let's look at this question through the lens of practicality.

- Transplants. You start with transplants because it is way easier or because you only need two of something, not fifty.
- Seeds. You start with seeds because there is no other choice. Either those items are available only as seeds, in which case your decision is preordained, or they're so new that transplants are hard to find. You'll want seeds for these: beans, beets, carrots, peas, radishes, and nasturtiums.

There are, of course, a few gray areas. For instance, you may well find transplants of peas for sale in early spring, but in my experience they don't transplant well, and besides, the seeds grow so quickly I can't see any reason to take a chance.

By and large, my strong suggestion for container gardeners, especially if they are new, is this: ***whenever possible, start with transplants***. This is especially true for cool-season plants in early spring: let someone with a professional greenhouse do the hard work of getting them established.

The tricky part of all this is balancing availability with timing, what I call the Gardener's Catch-22. It goes like this: you decide to wait for transplants, but they never appear, and by then it's too late to start seeds, even if you could find them this late. Fortunately, there is a way out of this dilemma, but it involves a little extra planning:

1. Buy seeds early, start them indoors, and move them out to their container spot when frost days are over.
2. Check with the nursery when the first cool-season transplants begin to show up. Do they expect to get in any of that thing you're interested in?
3. If so, plan to buy a few nursery transplants as soon as you can. Grow both, see which performs better.
4. If not, ask whether they will be offering something else very similar. Give that one a try.

This requires getting comfortable with the idea of double-duty planning. It also requires remembering to order the seeds in December or January, before most of the world is even thinking about gardens. Don't dawdle. We have just seen most of those mail-order houses run out of stock two years in a row (2020 and 2021).

There is an alternative—more expensive, less certain, but simpler. *Some* mail-order seed companies also sell *some* transplants by mail. If you find just what you're looking for, put in your transplant order as soon as possible. This is more expensive, of course, and you can't count on finding a mail-order supplier. If I were you I'd still order seeds as backup.

Where to Get Help

Does this all sound like a lot of work? It's actually easier than you think, because you don't have to do it all at once, and because the individual steps are kind of fun. There are probably millions of your fellow gardeners who can't think of anything they'd rather do on a Sunday afternoon in January than browse through this year's new seed catalogs. OK, almost anything.

And always remember this: you are not on your own. There are multiple avenues for help, on all manner of questions large and small. Here's a list of sources, starting with your biggest ally.

A friendly neighbor who's a longtime gardener. Make this person your new best friend. It won't be hard; gardeners are the nicest, friendliest people on earth. Just go over and introduce yourself; I guarantee you will be welcomed. You will probably find yourself turning to your neighbor for myriad questions in the beginning, but the key reason I'm pointing you in that direction is the ultralocal nature of what they can help with. You can find a wheelbarrow load of accurate general information about plants and techniques elsewhere, but only someone who lives nearby can fill you in on the peculiarities of the microclimate you'll deal with or warn you about a certain fungal disease that tore through the neighborhood last year. Bake a lot of brownies, and keep your questions at the ready.

A local garden group. Look for either a garden club or an online group for your town or neighborhood. This is essentially the same concept as your close neighbor, expanded just a little. If you learn about a neighborhood garden club, ask if you can come as a guest. I can't imagine you'll get anything other than an enthusiastic *yes*. If their meetings aren't at a convenient time for you, search for an online community of gardeners, the sort where members post questions and other members contribute answers. Here again, the more specific the geography, the better.

The nursery in your town that everyone recommends. Ask your neighbor, your garden club, or your online community what nursery is their favorite. You'll get various answers, of course, but I have a hunch one place will get the most mentions. Make it your business to get to know that one. That's where you are likely to find a wide range of plants; after all your research, it will be great to see those plants in person. And that's also where you'll find knowledgeable staff. They know the growing conditions in your town, they know what plants they have in stock and what's coming in soon, and they can steer you toward good choices for your sunny or shady garden area. Some can even diagnose a problem if you bring in a leaf from a troubled plant. If you need any of that kind of assistance, try to visit at a time when they're less busy.

County extension offices and Master Gardeners. Thanks to two acts of Congress in the nineteenth century, every state in the union has one (sometimes more than one) land grant college or university. The name comes from the terms of the deal. Each institution was *granted* a significant block of federal *land*; in return, they pledged to develop new educational and research programs in agriculture and engineering and to expand access for students of all income levels. In 1914, the US Department of Agriculture, seeing a need to make that academic strength readily available at the local level, created a system of county extension offices in partnership with the land grant colleges and universities; the universities serve as a kind of central base, with *extensions* out toward all the counties in that state. Today, every county in the country has an extension office, staffed with local experts who combine strong academic backgrounds with practical, on-the-ground knowledge to help farmers, small-business owners, and gardeners. Incidentally, John Fech, whose story you read in chapter 1, is a horticulture educator at the University of

I was born in September 1937, in Queens, New York, but spent most of my childhood in Florida—first Pensacola, then Jacksonville. My father was a civil service technician at the naval base, repairing airplanes and later helicopters. He was in the reserves and desperately wanted to go to war, but he was an essential worker and father of three, and the government kept turning him down for active service. Thank heavens.

We arrived in Pensacola in the summer of 1942, just in time for hurricane season. I came to believe that men in uniform and planes roaring overhead and yards totally given over to vegetable crops were a normal way of life. I planted my first green bean that year and tended it with a four-year-old's great fascination.

In the garden we grew peas, beans, corn, potatoes, tomatoes, peppers, cucumbers, radishes, and onions. Everything was planted in raised rows because in our section of Jacksonville we could dig a hole in the sandy soil and it would immediately fill with water! We all worked in the garden, weeding, watering, gathering, and then cleaning, removing the peas and limas from their pods, stringing and snapping the green beans, scrubbing the potatoes, and husking the corn. Then my mother discovered A&P frozen foods. My job, besides planting beans and peas, was mostly weeding. I didn't really mind weeding (still don't), but I hated the heat (still do).

With war rationing, many things were hard to come by, and families in our area developed a robust system of bartering. I remember galvanized tubs appearing, turned upside down over live chickens or filled with ice water and live crabs or shrimp. You don't want a description of how the chickens got from tub to table. Some neighbors gave haircuts or shaves, and my mother, who had been trained as a professional seamstress, made wedding and graduation gowns and turned cuffs and collars on uniform shirts. I think our garden excess was a part of that process. Those were fateful days for our country and desperate days for my parents. We all made it.

In 1972 my husband and I bought our first home and I proceeded to dig a garden for vegetables. With advice from the neighbors and pamphlets from the local extension service, I learned to garden in the muggy heat and red clay of Virginia. The memory of the Victory Gardens was so deep and so powerful that I couldn't imagine a backyard without a vegetable garden.

Marilyn Clark
Eden Prairie, Minnesota

Nebraska's Extension Center. As well as a gifted storyteller, he is a walking example of the inestimable value of extension offices and their dedicated staff members.

Throughout the year your county extension folks offer special classes, workshops, and a range of other community events that are worth keeping an eye out for. Online resources too: I'm willing to bet that they (or the statewide office) offer a free monthly newsletter, maybe something like "What to Do in Your Garden This Month." But for home gardeners, especially new gardeners, the most significant function is probably their sponsorship of the Master Gardener program.

The program varies slightly from state to state, but the essence is this: an intensive educational program of online and in-person classes, lab sessions, and hands-on workshops; all textbooks and other materials are included in the fee, which at the time of this writing is in the general neighborhood of $200. Many groups offer scholarships. Those who complete the classes, pass a final exam, and document a minimum number of volunteer hours (typically ranging from forty to seventy) are certified as Master Gardeners; that certification is good for one year.

It's that volunteer commitment that is so vital to home gardeners. The next time you're at a neighborhood outdoor fair and see a table offering "Ask Me About Your Garden," those are Master Gardeners sitting there. If your local farmers market has a rotating community table, I'll bet that once a month you'll find Master Gardeners there, eager to help. If your children come home from school talking about a nice lady who showed them about gardening, they just met a Master Gardener. And whenever you call your county extension office for garden information, the chances are you're going to be talking to a Master Gardener volunteer. Don't feel shy about asking what you think is a silly question; I guarantee you they have heard it all and then some, and they are invariably gracious. Gardeners are the nicest people, remember.

Mail-order gardening companies. Many of these companies, as we've noted, are using their online catalogs to give us deeply detailed information about each and every product, far more than would be feasible in the print catalogs. The scope of that information is truly amazing: sun and water requirements; the ideal growing season; the timing for planting, including whether you should plant seeds directly into your container (known as *direct sowing*) or start them indoors early; how deep to plant and how far apart; how long before germination; how long before ready for harvest; maybe even recipes and cooking tips. All this, and more besides, is at your disposal completely free.

In Summary, Then . . .

Here's a recap of your step-by-step planning:

- Right now, start paying attention to sun/shade patterns. You absolutely must have a clear idea of how much sunshine your potential garden area gets, because all other decisions flow from this.
- Then, start building your wish list. It's OK to start with broad categories, like "tomatoes, beans, lettuce, herbs."
- At the same time, start identifying all the places and people you can call on for help.
- Then, from their advice and your own research, compare your wish list against the reality of your space and your weather patterns. Reevaluate as necessary, and start the process of narrowing down general categories to specific varieties. Some items may have to come off altogether, but many of the items you're interested in will have special varieties developed just for containers. Your job now is to find them.
- Dig deeper into those special container varieties; look for things that fit your space/climate realities *and* your wish list. And identify alternatives, just in case you can't find your first choices. Be ready to be surprised at the possibilities.

Now, where are you on the calendar?

- If it's December or January or February and you're planning to buy seeds via mail order, get busy. After that, it's probably too late.
- If it's early spring, take your wish list to your favorite nursery and pick up transplants of cool-season vegetables.
- If it's mid-spring and you have seeds of warm-season plants that you need to start indoors, get busy.
- If it's late spring or early summer (depending on your local climate patterns), take your wish list to the nursery and scout out warm-season transplants.
- If you're starting this whole process in late summer or fall, you'll almost certainly be limited to planting cool-season plants for the fall, either seeds or any transplants you can find. Otherwise, use the time to research and dream for next year.

CHAPTER 3

Gear Up!

Congratulations! All the careful planning for your new garden is done: Dream. Doodle. Assess the space. Research the plants. Reevaluate the dream. Doodle some more. Decide, but allow for surprises. You did it all, and now you're ready for the next phase. Which is not planting. Not yet.

Before you can actually put plants in the soil, I recommend you gather all the tools, equipment, and supplies you're going to need. That's the goal of this chapter: to help you get the physical gear in place, all decisions made, and shopping done. I know, you're eager to get going. But a lot of what we talk about in this chapter can be done way ahead of the actual planting time, when you can't be gardening anyway. And then you're free for all the fun stuff when the weather gods smile.

Under two big categories, which I have labeled "Hardware" and "Software," I will introduce you to the supplies you will need. I'll also show you how to craft homemade versions that can ease your budget, and point out things you can do without. I'd much rather see you put your money into beautiful plants. You'll also find ideas for making the most of your space and keeping it clutter-free and handsome.

Hardware

If you're starting from zero, these are the pieces of equipment you need to acquire. Many are available online, but if you're able to do so, I encourage you to spend some time looking at the real thing in person. Browse the offerings at a large garden center or big-box home improvement store to get a broad sense of what's available and what the various options look like. Then shop where your heart and your wallet lead you. Maybe that same garden center, if you've fallen in love with some gorgeous pots, for example.

Plenty of other possibilities abound. I, known in my family as the Yard Sale Queen, love picking up beautiful things at estate sales and DIY possibilities at my favorite thrift store. Or pay close attention when visiting someone else's garden and ask where they found their things; we all borrow one another's best ideas. The point is, you'll probably be living with these items for years; make sure you choose something you really like.

Commercial containers, left to right: Fabric grow bag. Round ceramic. Tall plastic. Clay in traditional shape (showing that diameter and height are the same). Molded plastic. Another grow bag. Wooden barrel in rear.

CONTAINERS

The range of options is wide, so I'll describe the pluses and minuses of the most common types (see illustration above) and let you decide which works best for you. I am going to give you a strong recommendation, though: whichever style or material you end up with, make sure you acquire at least one very large container, the largest you think you can reasonably manage once it is filled.

Something to keep in mind: By long tradition, container size is defined as the diameter across the top expressed in inches. A ten-inch pot is one that measures ten inches horizontally at the top. The most common container shape has slightly tapered sides, down to a somewhat narrower bottom; picture a classic clay pot. And in that classic shape, the pot is the same height, in inches, as its diameter at the top. (See illustration on page 78.) Today, of course, there are millions of variations to these general characteristics, but the naming protocol is still the same: the diameter at the top is the "size" of the container. This all comes into play (and fosters confusion) when the same measuring system is applied to plants. When a nursery places a newspaper ad for six-inch plants, it doesn't mean that the plants are six inches tall; it means that they are in six-inch pots, which roughly suggests a certain level of growth. That same system often shows up in the garden-center aisles, too, with signs directing us to the "four-inch herbs." Those *always* refer to the pots, not the plants.

THE GOLDEN RULE OF CONTAINERS

Every container you plant must, absolutely *must*, have drainage holes. No exceptions, no excuses, no backtalk.

Clay. Good: Classic good looks. Widely available in a range of sizes at reasonable prices. Material is porous, which allows air to pass through to the roots, which is a good thing. Bad: Relatively heavy and breakable. Soil dries quickly because water vapor passes out through the porous sides very readily. Worse, if water in the pores freezes deeply enough to expand, the whole thing will crack wide open.

Ceramic. Good: Elegant shapes and beautiful, rich colors not available in any other medium. Bad: Vulnerable to breakage. Expensive. Sometimes does not have drainage holes.

Plastic. Good: Long-lasting. Inexpensive. Widely available in broad range of sizes. Bad: Not especially handsome, although we are now seeing newer products in pretty colors instead of basic black.

Wood. Good: Made from a renewable resource. Has a definite "look," which may or may not fit with your other design decisions. Bad: Not as widely available as other options. Can be expensive. Will develop rot after several years of watering, unless it has been treated with a wood preservative, which you don't want anyway because some of them are toxic.

Fabric bags. Relatively new on the market, these are well worth knowing about. Good:

Easy to move around (many of them have built-in handles). Attractive, slightly rustic look. Fabric is porous by nature, so air moves freely in to the roots, a good thing. When not in use and not filled with soil, they can be folded up and stored away. Bad: Tend to last only a few years.

Your clever ideas. So far we've been talking about the kinds of containers you buy at the store. But that is not the total universe of possibilities. Thrift stores and yard sales are brimming with items just waiting for your special touch. Keep your eyes open and ask yourself, "Could I make a garden container out of that?" Just remember, whatever you're considering will need drainage holes. If you don't own a household drill, now's the time to make friends with someone who does.

Do-it-your-clever-self containers. Front row: A plastic bucket is easy to move around. Toy dump truck for a child's planting. Back row: Restaurant-size can (the can opener known as a *church key* quickly makes triangular drainage holes). Large bowl, plastic or wooden. Wooden or plastic storage box. You *must* add drainage holes to all!

To give you an example, here's a personal story. At my favorite thrift store I once found a large wooden salad bowl, wide but shallow like most are, with the outside surface painted bright red and three little round feet, also painted red. I drilled three drainage holes, which is easy to do with wood, filled it with nice potting soil, and planted a happy selection of salad greens. Going outside to harvest salad makings from my salad bowl always made me smile.

TOOLS

Good news here: you need fewer tools than you might imagine. Some of them can come from the thrift store as easily as the big garden center, and only one of them, in my humble opinion, is worth spending serious money on. Here's your rundown. (See illustration here.)

Left to right, front row: gloves, pruners, cultivator, trowel, plastic write-upon plant labels. Rear: small knife, bucket, watering can.

Trowel. You need it for digging holes so you can slip in your transplants or gently remove plants that have passed their season. As long as you don't try to use it to lever up a concrete block, a modestly priced version will do just fine.

Cultivator. Looks like a giant fork with the tines bent down ninety degrees. Useful for fluffing up the top layer of soil in your container when you're getting ready for planting.

Hand pruner. This is the one you should budget for, in the range of thirty to fifty dollars. A good hand pruner will last you many years; a cheap one will drive you mad. When you're looking at possibilities, remember that in all likelihood you will not be pruning anything large and heavy; so products that emphasize their clean cuts on tree limbs, for example, are probably more than you need.

Watering can. Essential and wondrously inexpensive. Or you can make a nifty DIY version from plastic milk jugs. (See illustration below.)

Gloves. Serious garden gloves will protect your hands from sharp thorns, broken pottery, and giant slugs. For more run-of-the-mill tasks like weeding and light pruning, you may decide you don't really need them. Again, a very personal decision, but over the years I've been perfectly happy with a succession of inexpensive gloves that I don't mind tossing when they get unfixably dirty. Then, for chores that will involve lots of soil mixing and lots of water, where I'm not worried about pointy hazards but know my hands will end up a muddy mess, I use a pair of disposable plastic gloves intended for one-time use on household cleaning projects—the really thin, clear vinyl ones, not the up-to-your-elbow rubber ones meant for dishwashing.

Additional helpful items. Now, some other things you will find very useful but don't need to spend money on because you already have them lying around, or at least not much money (hello, thrift stores).

Household items make no-cost gardening tools. Front row: plastic clamshell container becomes a miniature greenhouse; serrated knife works like a small handsaw; barbecue fork for reaching deep into a container; chopsticks for many tasks. Back row: three ways to use plastic milk jugs.

- **A small, sharp knife** (from the kitchen) will come in handy for those tasks that the pruners aren't quite right for.
- **A big bucket** is handy in so many ways: carry water to the plants if your hose won't quite reach, move along with you as you do pruning or weeding, and much more.
- **A serrated knife** will actually function like a small saw, if you ever need to cut through a dense mass of roots or something like that.
- **Dishpan** or something similar. A good-size container, more broad than deep, like a square dishpan or a big storage container for which no one knows where the lid is. You'll find it handy for mixing potting soil with additives like slow-release fertilizer beads. Doesn't have to be pretty.
- **Manicure scissors** or a pair of those little clippers that seamstresses use to snip threads. Handy for tasks where you need to get to an exact spot on a plant and make a careful cut without accidentally cutting something else, such as snipping off dead or dying flowers that are still on the plant (a process with the gruesome name *deadheading*), or pruning plants with soft stems, like basil.
- **Wooden chopsticks** offer lots of uses, starting with making tiny, precise holes for small seeds you want in a particular spot. Your county health department forbids restaurants to reuse them, so you should feel completely free to take yours home with you after your meal.
- **Plastic clamshell-style containers** that fresh produce is packaged in. If you plan to start seeds early indoors, these make perfect miniature greenhouses. You can write on the outsides with markers to record what you planted and when (see illustration on page 75).
- **Plastic milk jugs** are made of material that cuts very readily with a sharp paring knife, so you can create several handy items by carving away unnecessary parts (see illustration on page 52).
    - ***From half-gallon size:*** Cut off a big hunk at the top but leave the handle intact. This is now a handy carrier for the garden tools you use frequently; keeps them all in one compact space, ready to grab and go.
    - ***From gallon size:*** Cut away the bottom entirely, so it looks almost normal, except that there is no bottom, just open space. Use to cover delicate plants threated by a cold snap. The cut edge is sharp enough that it's easy to scrunch the jug down into the soil, to secure it in position. Leave the cap on for complete weather

protection; remove it (but reserve) for air circulation during the day if temperature warms up. You have just created a perfectly workable substitute for a fancy garden cloche, which retails for thirty dollars or more.

- ***From gallon size:*** Cut away a big hunk from the top, down to the midriff, but leave the handle intact. Instant, and free, watering can.

Pat's story reminds us that there are some types of equipment that today's container gardeners won't have to worry about.

My father grew up in Ellis County, Texas, during the Depression, and a big vegetable garden was part of his life from childhood. The family lived on about half an acre in the small Texas town of Italy (if you want to call it a town), and the whole backyard, all the way up to the back of the house, was a huge garden. No lawn, just vegetables. In that same space they also kept chickens and two horses. My grandfather was using one of the horses to plow the garden space when he suffered a heart attack and dropped dead on the spot. The horse just stood there until someone came outside and found him.

But that tragedy didn't damage my father's love of gardening. After he returned home after serving in North Africa and Italy during World War II, he and my mom moved to Dallas and bought a house with a large backyard. Dad immediately created a vegetable garden, where he grew just about every kind of vegetable and gave away the extras to neighbors. He never officially "exercised," but he was always fit because he worked in his garden nearly every day, weather permitting. He lived to ninety-five but never plowed with a horse, only a gas-powered tiller.

Pat Pape
Argyle, Texas

Software

Not *that* kind of software; I'm talking about supplies you will need to have on hand when you're ready to start planting, things with a softer physical character.

POTTING SOIL

The simplest strategy, and the most efficient, is to purchase bags of potting soil from the garden center. Look for products labeled *potting soil* or *container soil*, something like that. Technically, scientifically, it's not soil at all. It's a carefully balanced mix of finely shredded peat moss and tree bark, with one other ingredient added to keep the other particles from clumping together into a dense mass. That third ingredient might be perlite, which looks like tiny white rocks but is actually fine bits of naturally occurring volcanic glass; or vermiculite, a natural mineral that shows up as thin flakes with a lightly mirrored surface; some mixes have both.

You may also find specialty products with extra ingredients like fertilizer beads or water gels (explained below); you may prefer to add those things on your own instead. Other special blends are those labeled as ideal for a certain type of plant, such as vegetables or rhododendrons or lawn grass. What's going on there is that they have extra ingredients to support those plant categories: balanced fertilizer for vegetables, something highly acidic for rhodies, and lots of nitrogen for lawns. I suggest you ignore all that and just get a nice all-purpose potting soil. To find a good one, try lifting the largest bag you see; if you're surprised by how light it is, buy that one. Light is good. It means that the mixture contains a good measure of that critical third ingredient that will keep the soil loose and easy to work with, and which is itself very lightweight.

If you are a dedicated organic gardener, you may prefer to search out organic soil. Fortunately, with such increased consumer interest, this is much easier than it was just a few years ago. The suppliers tend to be regional rather than national, so this is another case where your local gardening connections are your best resource for information.

Another item you may well find on the same

POTTING SOIL TIP

Scan the shelves for a large bag of all-purpose potting soil. Try to pick it up. If it's way lighter than you thought it would be, that's the one you want.

retail shelves is topsoil; leave it there. Topsoil is dense, hard to work with, and not especially nutrient-rich. And the one thing I definitely want you *not* to do is dig up dirt from someone's yard; you never know what weed seeds or nasty pathogens can be lurking.

So watch the sales, pick up a few bags of all-purpose potting soil from a company you've heard of or one suggested by your garden network, and you're set.

FERTILIZER

You will frequently hear people say, when they talk about fertilizing, that they are "feeding" their plants, but the truth is, in one of those delightful marvels of science, plants manufacture their own food through the process of photosynthesis that you learned about in middle school. That manufacturing process depends on sunshine (you already knew that, from chapter 2) and on certain chemical nutrients found in healthy soil. Augmenting those nutrients if they are weak, or replacing them if they are totally missing, is the purpose of fertilizers.

Critical nutrients. Plants need three main nutrients: nitrogen, phosphorus, and potassium—chemical symbols N, P, and K, respectively. To a lesser degree they also need other nutrients and trace minerals, but those three are essential. Nitrogen stimulates growth of leaves and their green chlorophyll (needed for photosynthesis). Phosphorus provides growth energy for new plants, helping them develop flowers. Potassium builds strong root systems and promotes overall hardiness. So as a broad generalization, we can say that . . .

- Nitrogen promotes leaf growth.
- Phosphorus promotes flower growth, and thus the fruits (vegetables) that come after the flowers.
- Potassium promotes root growth.

You want a fertilizer that matches what your plants need. Once upon a time, manufacturers displayed the relative amounts of the three main nutrients as big numbers on the front of the package, like 8–17–12 (the percentages of N, P, and K—always in that order—in the product). But in recent years most manufacturers have moved that information to the back of the package, in small type, and instead label their various products by the type of plants each is formulated for. Like this: "Tomato Plant Food." Or "Turf Builder Lawn Food." Or "Rhododendron and Azalea Food."

So take the time to read the label. Also, get familiar with these two terms often used to describe fertilizers, because you're likely to come across them in your research. A *complete* fertilizer has some portion of all three main nutrients. A *balanced* fertilizer contains those three main nutrients in equal or nearly equal proportions, such as 10–10–10 or 8–10–7.

Now, I don't for one minute suggest that you purchase several different fertilizers to accommodate the various things you'll be planting. For most of you, most of the time, a complete fertilizer in which the middle number—phosphorus—is higher than the other two will serve you well. Phosphorus, remember, promotes flower growth, and many of our vegetables start with a flower. So something in the ratio of 5–10–5 would work fine as an all-purpose fertilizer. *Ratio*, not necessarily absolute numbers. On the other hand, if you are quite sure your garden will have more green leafy veggies than anything else, pick a formulation in which the first number (nitrogen) is more or less equal to the second.

Physical form. One other thing to get acquainted with: the physical form of the fertilizer itself. You can buy fertilizer as a concentrated liquid, a powder, or a short spike made of compacted, slow-release fertilizer designed to be stuck in the soil. You can also choose slow-release beads, about the size of BBs, with a coating that dissolves in water, so that the fertilizer is gradually released over time. Or slow-release granules in a container with holes in the dispenser top, allowing you to shake the small granules around existing plants. Both are very convenient.

There is an issue of timing to keep in mind. Most slow-release granules are meant to be worked into the potting soil before you add any plants or seeds. To add more fertilizer during the main growing season (a good idea, by the way), you'll need something else: one of the shaker types or one of the types that you first dissolve in water. I think you'll find this is not really cumbersome, just something to be aware of as you're planning your shopping list.

Organic versus inorganic. Organic fertilizer is made from something that was once alive; inorganic fertilizer is manufactured using synthetic materials. Bonemeal (from crushed animal bones), fish fertilizer (from decomposed dead fish), packaged manure (from you-know-what)—these are all examples of organic fertilizers. Commercial manufacturers also produce organic fertilizers, proudly labeled as such, with a combination of ingredients. Inorganic fertilizers are made from synthetic versions of the same essential nutrients; the constituent chemicals are the same, and so is the effect on your plants.

The decision is largely a matter of principle, something I support unconditionally. But if you're undecided, here are a few things to keep in mind. As a general rule, organic fertilizers tend to have lower levels of the main nutrients and release those nutrients more slowly. Measure for measure, inorganic fertilizers will give you faster results. On the other hand, organic fertilizers usually stay active in the soil longer. One of life's pesky trade-offs. Also important for vegetable gardeners: speaking very generally, most of the formulas of organic fertilizer products are proportionately low in phosphorus, which is the one nutrient we need the most. Don't choose a product simply because the package says *organic*; study the label for the breakdown of ingredients.

And please don't get sidetracked by emotionally loaded terms. Organic doesn't automatically mean safe, and inorganic doesn't automatically mean unhealthy. *Natural* is not the same as *nontoxic*. The natural world is full of things untouched by human hands that are quite dangerous, even deadly. And *chemical* is not a dirty word. The entire universe is made up of chemicals—starting with water.

Compost. A special form of organic fertilizer I heartily recommend is compost, which is the end result of carefully tended decomposition of organic materials such as vegetable scraps and plant trimmings. In finished form it's dark brown and crumbly, but vegetable gardeners think of it as solid gold. Compost provides the key nutrients we look for, but so much more. It also greatly improves soil structure and enhances the soil's capacity for holding water, extremely good news for container gardeners.

Many home gardeners make their own compost, and if your new gardening pals offer you a bucketful, by all means say yes. It may not be practical for you to do yourself—the physical setup takes space, plus where are apartment dwellers going to get lawn trimmings?—but fortunately you can purchase bags of commercial compost at your favorite garden center. This is another item you want to fold into the soil before you start planting; then, as the growing season moves into high gear, add a layer on top of the potting soil and work it in gently.

WATER GELS

This product is a big help toward easing a container gardener's biggest frustration: the need for frequent watering. These tiny crystals of a space-age polymer gel absorb many times their size and weight in water and then gradually release it over time, all the way down to the original crystal size. Next time you water, the process starts all over again. This will not totally relieve you of the duty to pay attention to watering, but it will help. This is another item you work into the potting soil before you add any plants or seeds. Aren't you glad you found that big dishpan?

I want to tell you about my father's Victory Garden. He did not grow up in a farming family; his mother was a teacher and his father an engineer. My dad became a civil engineer himself. I think we can say he was not a natural gardener.

Starting in 1932, my father ran the brand-new wastewater treatment works for Toledo, Ohio. As you probably know, the by-product of that process is quite good as a "natural" fertilizer, and in this case the sludge contained seeds from the wastewater produced by the nearby Heinz canning plant. My father noticed that the dried sludge produced quite viable tomato plants, and he decided to explore the possibilities. So with an eye toward the commercialization of dried sludge fertilizer, he carefully laid out a Victory Garden plot in our front lawn at home.

The goal was to produce tomatoes of exceptional quality, so my father the engineer started by consulting with expert tomato growers. He purchased four dozen small plants for the new plot and tended them faithfully as they grew into robust plants. The experts had told him that he should remove the new sprouts that developed in the joints between branches and the main stem, or else the flowers would not form to produce fruit. So every week, Dad faithfully removed the "strength-sappers" from the forty-eight plants.

With plenty of water and lots of sludge fertilizer, the plants grew quickly, easily reaching twelve feet tall. Dad had to buy a stepladder to reach those "pesky" growths. But no tomatoes. After two months of fruitless plants, he finally cut them all down. He had learned, at long last, that those sprouts he had so carefully trimmed were the flower-producing parts of the healthy plants.

Dad took a lot of kidding when his colleagues heard the story of the "miracle" sludge fertilizer and the fruitless tomato Victory Garden, and to my knowledge never tried gardening again. Until . . .

Long after World War II was over and Dad and Mother had moved to Florida, he tried his hand at growing a lemon tree. Unfortunately, the house A/C drain dripped onto the tree roots and killled it!

Louise Bankey
Portland, Oregon

The material itself is completely safe for plants, but you don't want your kiddos to sample it. You also don't want to work with it around your kitchen sink; any you spill into the drain could totally clog it up the first time you turn on the tap.

Making the Most of Your Space

When you have limited garden space to work with, you have to be extremely clever about how you use it. There are two ways to approach this: thoughtful physical arrangements that maximize floor space, and staged planting in which one container has multiple crops as the seasons change so that something is always growing. I hope you'll do both.

SPACE-SAVING CONSTRUCTIONS

I'm going to assume that your garden space is pretty small—a balcony maybe, or a small patio or deck. If I'm wrong, you won't have to rely so completely on these space-saving ideas but they might still come in handy for you.

Container pyramids. I came up with this idea out of desperation; my little patio is, well, little. This pyramid gives me a lot of planting space but has a quite small footprint. And it's dead simple to put together. (See illustration below.)

Space-saving pyramid. Three pots of exact same height, with rims touching in the center and fourth pot on top—lots of planting space in a small footprint.

Start with three containers of a pretty good size—at least twelve inches in diameter across the top; sixteen or eighteen is better. Arrange them in a tight triangle; the rims should be touching in the center. Then, where the three meet, position a fourth container on the rims; it should be somewhat smaller. Now you have a tower that measured from any angle is no more than twenty-four inches wide (or thirty-two or thirty-six if you use the larger containers). Out of that tight bit of floor space you will get a *lot* of growing space.

This works only if the three bottom containers are exactly the same height. And aesthetically it's more pleasing if they are the same color and the same material. So it's much simpler if you purchase them at the same time. The top container can also be in that same family, for a cohesive look, or a contrasting color if you prefer.

Baker's rack. This idea comes from Jerri Austin Marler, of Corvallis, Oregon. (See illustration on this page.) A baker's rack like this is typically used for extra storage in home kitchens, but Jerri realized one would make an excellent addition to her patio garden, where it takes up very little floor space but provides several levels of shelving for containers. The

An elegant baker's rack holds many containers in a space-saving vertical format.

same could be said of any type of storage rack, of course, like the utilitarian shelves someone might have in their garage: they do the job, but no one would call them beautiful. In contrast, the glory of this style of baker's rack is how lovely they are. Even when your containers are just getting started, that corner of your garden space will be a visual highlight.

To hold the weight of filled containers, the shelves need to be sturdy. And brand-new, high-quality pieces are not cheap. But the Yard Sale Queen reminds you that this is the sort of item that often turns up at estate sales. If you see one, grab it.

Trellises. Both my pyramid and Jerri's baker's rack are adaptations of a valuable theme that has multiple variations: growing plants vertically. If a plant can be encouraged to grow upward instead of outward, you will save a great deal of your floor space but at no loss of growing space. The classic method is a trellis of some type. Hardware stores, home improvement centers, and garden centers all have multiple possibilities. Take along your favorite handyperson and collect ideas for custom-made versions, in case none of the commercial items strikes your fancy.

Your next assignment is to find a way to anchor the trellis so it's not vulnerable to winds or even to falling over from the weight of whatever is growing on it. Sometimes a trellis has long "legs" at the bottom, which might be sufficient as an anchor if you can sink it deeply enough into your container. If your container is wooden or plastic, you can bolt the trellis in place with strong screws; it's easier to do when the container is empty.

Here's another homemade vertical variation: On the patio side of my dining room window, I zigzagged fishing line up and down from the top windowsill to the bottom, snuggled an outdoor storage cabinet up against the bottom sill, put a horizontal planter on top of the cabinet, and filled it with seeds of sugar snap peas, which grabbed onto the fishing line as they started to grow. Viewed from the inside looking out, I had a beautiful green curtain; from the outside, I had delicious peas to harvest. In non-pea season, the fishing line is virtually invisible.

MULTIPLE PLANTINGS IN THE SAME CONTAINER

The other way to get the most out of your limited space is to make sure something is always growing in every square inch, using the technique called *succession planting.* In essence, it involves rotating seasonal plants in and out of the same planter so there is always something producing heartily and a minimum of bare soil in between. My personal mantra is "no bare dirt."

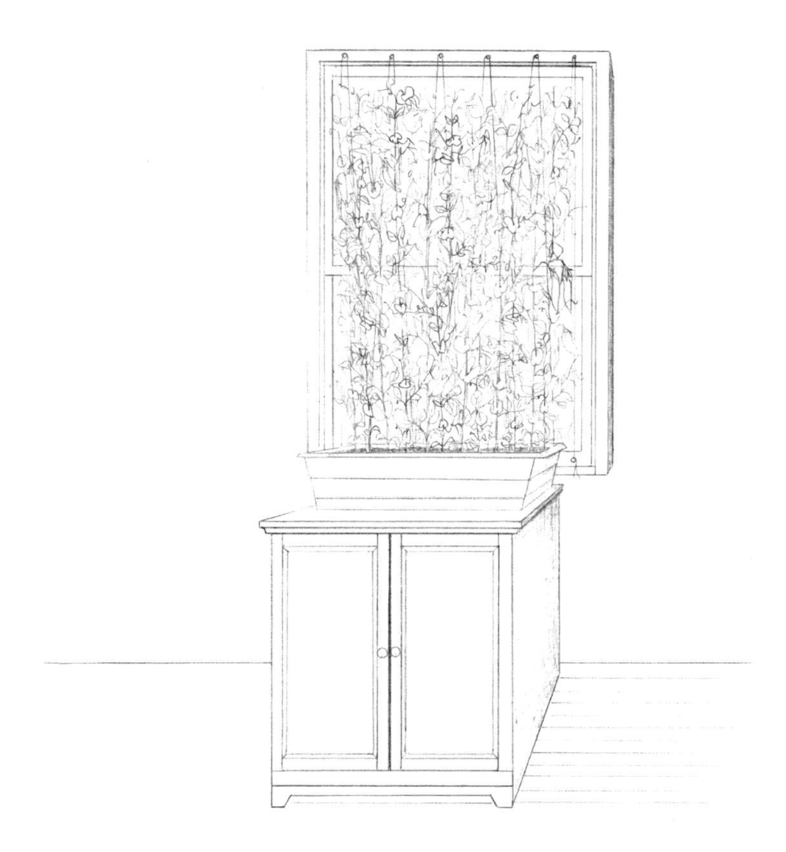

Sugar snap peas, growing in horizontal container on top of garden cabinet, climb a zigzag trellis of fishing line on Maggie's window.

Here's the basic pattern: Put in some cool-season plants as early as possible. When the temperature starts to climb and those plants begin heading downhill, add in some small transplants (or even seeds) of warm-season plants. When the cool-season things are totally kaput, pull them out (or cut away at the soil line if they don't come easily). By then the weather has warmed up significantly and the warm-season plants are about to burst into a serious growth spurt, which they now have room for. Then, as the summer nears the end, have another round of cool-season plants (or seeds) ready to slip in when the warm-season plants are done.

THE GOLDEN RULE OF CONTAINER GARDENING

No bare dirt.

And throughout the year, have a few things on hand to fill in any tiny bare spots. Like radishes, which grow quickly from seed; or carrots, which will take a little longer but look so pretty while growing; both of them need just a tiny bit of soil space. Or my personal "no bare dirt" secret weapon: green garlic, described here.

GREEN GARLIC

It's what we call the spiky green foliage of garlic plants. Super easy to grow, very versatile in cooking, and best of all, takes up very little growing space. (Note: This is not the same as growing garlic for new heads. That requires a long growing period, with rather fussy timing, and the green shoots are left on the plant to die down.)

Here's how: Start with a head of garlic, straight from the supermarket. Break off several individual cloves. You'll notice that each clove has one squared-off end (that's the bottom) and one pointy end (that's the top).

Now, in that tiny bit of container space with nothing growing in it, take a table knife, slide it down into the soil about two inches, and wiggle it back and forth a few times. Your goal is to create a miniature canyon, deep but narrow. It should be approximately twice as deep as a clove of garlic is long. Slip one clove down into the canyon, pointy end up, and cover it with soil.

Slip individual garlic cloves into any tiny bare spot in your container, and before long you'll be harvesting delicious green foliage that tastes like a garlic version of green onion and takes up almost no space.

And that's it. Really. You have nothing more to do except watch it grow.

In a couple of weeks you'll see tiny, green, pointy spears, which will grow to long, strappy foliage that looks very much like the green parts of green onions or scallions. Their taste is garlicky but not overwhelming, and you can use them the same as scallions in the kitchen, either in cooking or as garnishes. Snip off one or two for tonight's dinner, and leave the rest for another day. (And the ones you cut will regrow.) Best of all, while growing they take up a soil space smaller than a quarter and about as much vertical space as chopsticks. You may find yourself searching around for other small empty spots in which to tuck in more cloves.

Clutter Control

You are going to produce a bountiful, beautiful garden; I know you are. But it's all too easy for clutter to accumulate—tools, partially used bags of soil, packages of fertilizer, not to mention all those nursery pots you don't know what to do with. If you don't have a garage, where will you put it all? And pretty soon that clutter can overwhelm your pleasure in your garden. What we're going to do is keep it from happening in the first place, by creating storage systems to corral the mess.

STORAGE SYSTEMS

There are many commercial products to choose from, including various renditions of the same basic concept: a storage cabinet made of weatherproof material, designed for outdoor use. They come in a wide range of sizes and configurations and a few different colors. If this is the route that makes sense for you, I would just offer one word of caution: choose a cabinet with doors, rather than a lift-up top; you will invariably want to keep stuff on the top, and then it's ridiculously inconvenient to get into the cabinet.

But that's not the only route. With your clever brain and a few standard items from your favorite big-box "everything" store, you can put together storage systems that just exactly fit your space and your personal style. Here are a few ideas.

Dual-purpose bench. A bench is for sitting on while you watch the birds or daydream, but it also provides a well-defined storage space underneath. So find a bench you like and measure the underneath space. Take those measurements to the big-box store and buy a couple of big boxes that will fit. You'll be happier if they are clear material, so you can see what's inside. Also make sure they latch securely, since they'll be holding things you don't want to get wet, like partially used bags of potting soil. And if your new bench is made of wood, give it rain protection with a careful coat of polyurethane on all surfaces, including the underside.

Clear plastic boxes, tucked underneath a bench, reduce visual clutter but keep supplies handy.

A beautiful box, repurposed. Keep an eye out for a beautiful large box, like a child's toy box, a blanket chest, or even a small cedar chest. When you find one, visualize it standing on end: Can you see that as a storage box for your garden tools? If so, snag it.

Maybe give it a nice paint job to enhance the color scheme you're developing, and if it's wooden, add a coat or two of polyurethane for weather protection. Depending on the size, you may also want to add an interior shelf, or a support strip under the side that is now the top, or both. This is a pretty simple job for your favorite handyperson; just have a big pan of chocolate chip cookies at the ready.

Here are two versions of the same basic idea. One has a horizontal orientation, the other vertical. Use whichever best fits the shape of your box and your overall space.

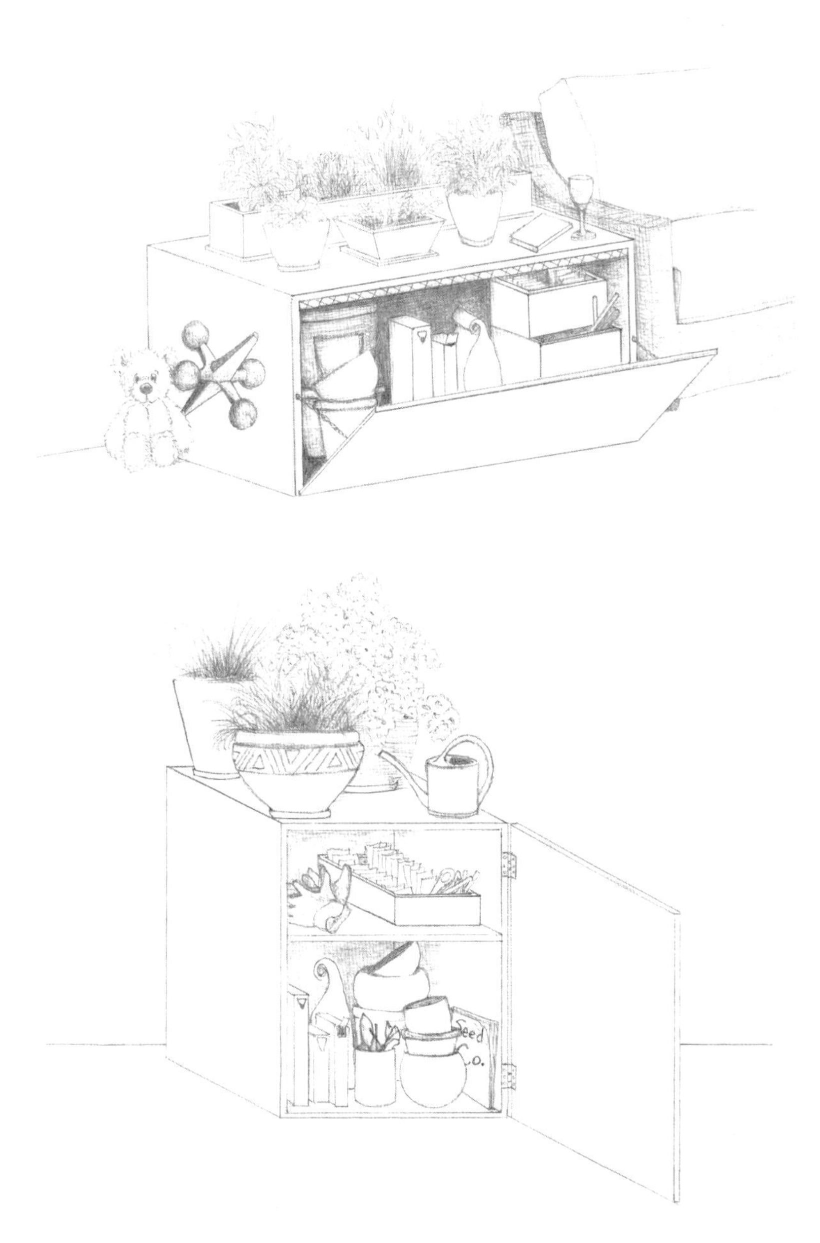

Wooden chests repurposed for garden storage.

VISUAL CLUTTER TOO

Even if you keep all the *stuff* out of sight, you can still end up with a cluttered look to your garden. More often than not, containers are the culprit. Which means they are also the solution.

You may not realize it at this early point, but the containers themselves are a big component of the visual impact of your garden. If they are a messy jumble, the whole look will suffer. If they are handsome and unified visually (shape, color, and/or material), they will greatly enhance your garden scheme.

If you're just starting your first garden, you're in good shape here. You can just buy matching containers—same material, same color, probably a few different sizes. You have to buy them anyway; I'm just asking you to think about choosing a cohesive look and stick with it. But if you have amassed or inherited a motley assortment of pots in different sizes, shapes, materials, and colors, it's time to get tough. First, take inventory. Dispose of the small pots, anything less than six inches in diameter. Save the rest, wash them carefully, and organize them by size and shape.

Do you still have a mishmash? Are they still ugly? Get thee to the paint store. Yep, regular house paint. Easy, inexpensive, and the results are amazing. You can use interior wall paint, but over time it may begin to flake off in spots, which is either a nuisance or a deliberate look, depending on your personal design style. To avoid that altogether, choose either exterior house paint or a spray paint formulated for outdoor use; one popular brand is Rust-Oleum.

What color? Up to you, but keep a couple of things in mind. Cool colors—blues, greens, purples—recede, so they will not visually overwhelm the plants; in fact they seem almost like neutral colors, forming a sophisticated foundation for the plants. Warm colors—red, yellow, orange—will jump out at you and demand your attention, more so than the plants themselves.

Also: Are there some nearby colors you might want to coordinate with? Perhaps the siding of your house, the trim on your windows and doors, or the cushions of your patio furniture? Or even elements of the view outside your garden, like the foliage of the nearby hedge or the lake in the distance.

PAINT YOUR POTS

To achieve a harmonious, peaceful look in your garden, paint all your containers the same color.

In Summary . . .

Even if it's not gardening season yet, you can get a big head start by browsing and collecting (and repurposing, if that's your style) the physical gear and materials you'll need when the time comes. It's a good way to fill the nongarden days, and it's also wonderfully satisfying. Through your careful planning, shopping, and crafting, you now have everything you need for your glorious garden.

Everything, that is, except for the seeds and the plants. That comes next.

CHAPTER 4

Designing, Planting, Nurturing

This is the part where you get your hands dirty.

Now is the time when gardeners get giddy with excitement, the garden centers stay open late, the weather forecast takes on critical significance, and all butterflies and bees in your neck of the woods send out the signal to their buddies. After all your careful planning and preparation, it's time for the real thing. This chapter is all about getting your garden going and thriving.

Success with Seeds

You will inevitably be planting some seeds, either because what you wish to grow doesn't come any other way or because you're in love with a new variety that isn't widely available yet as transplants. You will either put them straight into your container, known as *direct sowing*; or, if your climate patterns call for it, start them indoors ahead of time. Instructions for both are in the following pages.

Finding seeds is easy; finding very specific seeds may not be. In the springtime, every nursery, garden center, home improvement store, hardware store, general retailer, and fast-food restaurant (kidding on

YOU'LL NEED SEEDS FOR THESE

Beans. Beets. Carrots. Peas. Radishes. Nasturtiums.

that one) will have a seed rack or two or twenty. If you have a general idea of what you want, and you're not searching for a very specific variety, you'll have lots of choices, all good. But if you've been exploring the catalogs or websites of our many wonderful mail-order companies (which I hope you have), your range of choice is much wider and more fun. My only caution in that case is to place your mail orders in December or January; I know I'm repeating myself, but I don't want you to have to hear, "Sorry, all sold out."

If your seeds arrive before you're ready to plant, just keep them in a dry, cool cupboard. Don't open the package to admire the contents, unless you're splitting an order with a gardening buddy; in that case, reseal the opened packet as tightly as you can.

And that leads to another question: Should you save leftover seeds for next year? My recommendation is no. No matter how careful you are, some moisture from the air is going to find its way into that opened packet, and once that happens the seeds are pretty much ruined. In the grand scheme of things, the cost of a new packet of seeds is negligible.

SEEDS: TWO WAYS TO PLANT

A seed is a tiny miracle: it already contains everything the future plant needs to develop and grow into a pine tree or a holly bush or a butternut squash. The seed is dry, which is nature's way of holding it safely in limbo until the right weather cycle rolls around again. But once it absorbs a little moisture and the temperature is just right, the seed will burst open and start making roots and stems and leaves. It doesn't need fertilizer to do that. It doesn't even need to be in soil, and you couldn't stop it if you wanted to. So the first thing is, relax. Planting a seed isn't complicated; you learned how to do it in kindergarten.

Direct sowing. This means planting the seeds straight into its ultimate home in the soil, as is. To start, make sure the top layer of soil in your container is loosened up and very lightly damp. Use your cultivator or your fingers to break up any clumps and then go over everything again for good measure. Study the seed packet for instructions on how deep to plant the seeds and how far apart, and do that. Cover with damp soil, give the top a friendly pat, and you're done.

Starting indoors. This is what you do if you need to get a jump on your planting season.

I have to be honest—this is not exactly simple. Getting the timing right can be tricky, especially if the weather takes an unexpected turn in the middle of the process. And the little darlings from your windowsill can decide to go into shock when you move them outside unless you plan for a period of gradual adjustment (known as *hardening off*; see step 7 on the next page). On the other hand, you may not have a choice. And success is going to feel great.

There is more than one way to do this. Your gardening friends may have their own techniques to recommend, or you yourself may have done this in the past. All roads lead to the same goal, so if you're confident with a particular method, by all means stick with it. But if you've never done this, please consider using the system I describe here; I believe it will bring the best results with the fewest headaches.

PEAT PELLETS, STEP BY STEP

Peat pellets are made of finely shredded peat moss, encased in a thin mesh material, and compressed into a flat, round disc about the size of an Oreo cookie. When wet, they expand into a cylinder about three inches tall, into the top of which you deposit one (or more) seeds. As the seeds germinate and grow, the roots work their way out through the mesh. When the weather is right, you plant the entire bundle, mesh and all, straight into your container. Thus, no transplant shock, which is a huge advantage.

Peat pellets, from beginning to transplant-ready.

1. Get a supply of peat pellets from a retail garden center or mail-order company. They are not expensive, and they last indefinitely—as long as you keep them dry. You could split a package with other gardeners if you want to start small.
2. You'll need something waterproof to keep the pellets in while the seeds are growing. You can find purpose-designed commercial trays at the same retail sites, but simple DIY solutions are everywhere. Those clear plastic bins that originally held salad greens at the supermarket work well; you can write on the outsides with markers, noting what you planted. With their built-in tops, they function as miniature greenhouses. Just be sure the bin does not already have holes in the bottom; that would defeat the purpose.
3. In a large bowl, submerge the pellets in water (you may need to weight them down at first) until they swell to full size. Then transfer them to your tray.
4. Poke holes in the top of the pellet. A pencil works fine; this is also a good place to use your salvaged chopsticks. Insert seeds, one or two if large (beans, for example) or several if tiny.
5. Be patient. Check occasionally that the sides of the pellets are still damp; add more water at the bottom if not. If the emerging shoots all lean in one direction, toward your sunshine, rotate the container. Once the new green shoots bump into the top of the plastic bin, remove it or keep it propped open.
6. If your baby plants are getting so big that they start to flop over and it is still too cold outside, you might have to move them into intermediate pots that are still small enough to manage indoors. Or you can just say to yourself, *Dang it, I'm starting over.* That works too.
7. Once the weather decides to cooperate, you're ready for an important safety step: gradually acclimate the little plant from the warmth of your home to the outdoor climate, in a process called *hardening off.* You do it in stages, over the course of four or five days. First day, take the new plant outside for an hour; second day, two to three hours; and so on. By day 5, you're ready for the real thing.
8. Transplant the pellets into your container. Remember, you plant the entire thing as a unit: mesh covering, peat, root ball, and seedlings. Handle it just like all your other baby plants. Dig a hole, insert the transplant so that the top of the pellet is level with the soil line, water thoroughly, backfill the soil, give it a nice pat at the top.

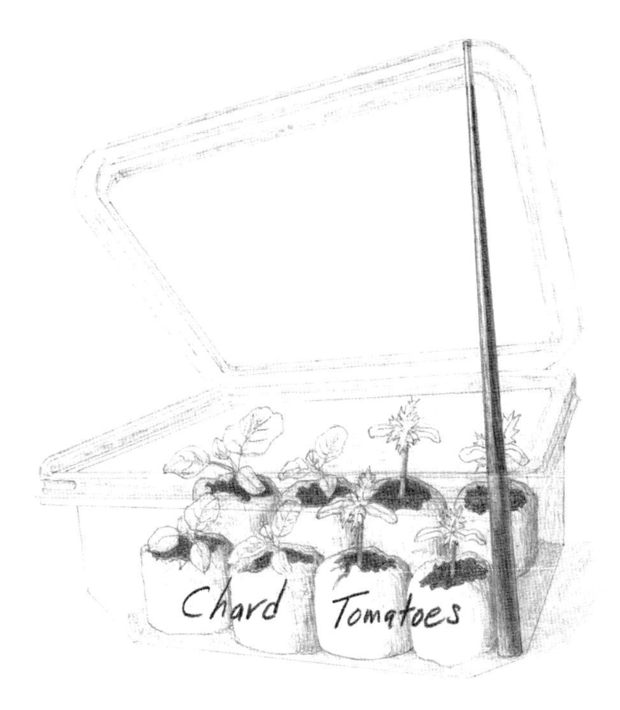

Recycle those large clamshell plastic bins from the grocery store into mini greenhouses. Bonus: you can write on the outsides with markers.

SEEDS: THINNING

You put two pea seeds into your peat pellet because you're a cautious soul. You plant a dozen carrot seeds in one spot because they're so tiny you can't separate them. You shake more radish seeds out of the packet than you meant to because the cat bumped your arm at the crucial moment. Or you plant what you think is just one beet seed but half a dozen seedlings emerge; guess what: that's how beets work.

In all those cases, and more, you now have a very important job to do. Remove the weaker seedlings so that the strongest ones have room to thrive. This is what we mean by *thinning*, and it applies both to seeds you started indoors and those you direct-seeded in the container. Pull the weaker ones up (or snip them off at the soil line, so as not to disturb the roots of the remainders) and discard. Yes, it will make you feel like a murderer. Do it anyway.

When several seeds germinate very close together, you must do some thinning. Choose the strongest-looking seedling and remove the others. Manicure scissors will help you get into just the right spot.

Success with Plants

Are you ready for the fun stuff? This is the point where you go shopping for new plants, spread them out, talk to them while they spend a day or two acclimating to your space, plant them into your beautiful containers, then stand back and smile.

PLANTS: WHERE TO BUY

Some sources are obvious: your favorite garden center or the garden section of a big home improvement store. Here I'd like to send a nod to the folks at your local independent nurseries; in my experience they are invariably savvier about local growing conditions and more willing to answer questions, and they deserve our support. If you have a shop like that in your town, spend some time there, and some of your budget; I think you'll be glad.

There are other sources, perhaps not quite so obvious.

- Many garden clubs and Master Gardener groups hold special events at the start of the gardening season. The festivities often include a fundraising plant sale, in which people bring in potted transplants from their own gardens, which means they can answer very specific questions. Check back with your information sources (review chapter 2) and make sure you're on their mailing lists for announcements of special events.

- Similarly, your extension office, either county or statewide, may offer a "welcome to spring" day-long gardening fair. Here again, extension staff or volunteers may staff a sale table of plants to raise money for the group. Often local nurseries and specialty growers are also invited to sell their new products.
- Keep an eye out the next time you visit your local farmers market. Many markets require vendors to sell only those items they themselves have grown, and quite often those who raise vegetables and herbs for the market will bring in small plants to sell early in the season, especially if they have more baby plants than their own fields can accommodate. I'm betting you will find many organic plants, and you'll get to talk to the actual human being who grew them and will be delighted to tell you about them. And here's something totally off the subject but very, very cool: there's a good chance that local chefs are standing right there beside you, looking for inspiration. One of Portland's famous chefs, studying the foliage of a new basil, was overheard asking the vendor, "Will you grow this for me?"

PLANTS: WHAT TO LOOK FOR

You want baby plants that are strong and healthy; that much is obvious—and pretty easy to spot. But I also want you to think in terms of beauty, of choosing specific plants of specific varieties that will delight you with their coloration and blend beautifully with the overall look of your garden. No reason you can't have both good health and good looks.

For best specimens, look for these attributes:

1. **Compact growth.** The plant already shows signs of a tight, compact structure. Avoid plants that are leggy, meaning those that have long, stretched-out stem sections between two sets of leaves. A plant is leggy because it has been desperately reaching for sunlight, which indicates to us that attention was not paid during its days in the grower's greenhouse. It's not a catastrophe, but if you're debating between two specimens of the same thing, choose the one that isn't leggy. (For a visual of legginess, see illustration on page 151.)

2. **No obvious signs of distress.** No dead or dying leaves. No weird discolorations. No holey leaves that might indicate insect problems. Overall, an impression of vigor rather than struggle.
3. **Too-early flowers or fruits.** It can look tempting, but resist. This plant has been in a greenhouse too long; it will struggle to acclimate to your outdoor space.
4. **Not root-bound.** This is probably the most common problem. The term refers to a tight bundle of roots on the plant, deep inside its little container where you can't see, and it means that plant has been in that small container too long. That tangle of roots needs a good haircut before you plant it; otherwise they will just continue growing around in a circle and the plant will never really take off.

   In the garden center, it's fairly easy to check. Pick up a couple of candidates, delicately tip out the plants and look at the root system on the bottom. One good clue: if one you pick up is quite light, this means there are more roots than soil; even if it was watered properly, it will still feel light because there isn't enough soil to absorb that water.

   It is, however, not necessarily a disaster. If you want that particular plant because it's great in every other respect, go ahead. Here's one spot where my advice is different from what you may hear elsewhere. I have found over many years of experimenting that it is not necessary to "gently untangle" that messy snarl of roots. If the root mass is really dense, I might soak it in water for a few minutes first. Then I just yank that snarl apart, cut off the worst clumps of knots, and it's ready to plant. The plant will grow more roots in a more reasonable sequence, thanking you all the while.

CHOOSING PLANTS FOR BEAUTY

Imagine yourself cruising the aisles of the garden center, wish list at the ready. Now you get to convert those wishes into real life.

Here's my tip: when shopping for a container designed around a certain theme or color pattern, try to purchase all those plants at the same place on the same trip. That way, you can use the shopping cart to pre-stage your combinations. Once you see them side by side, it will be instantly clear whether they blend harmoniously and enhance each other the way you envisioned. Or not. Just as quickly, you'll see if something looks not quite right next to something else, and it's easy to switch while surrounded with other choices.

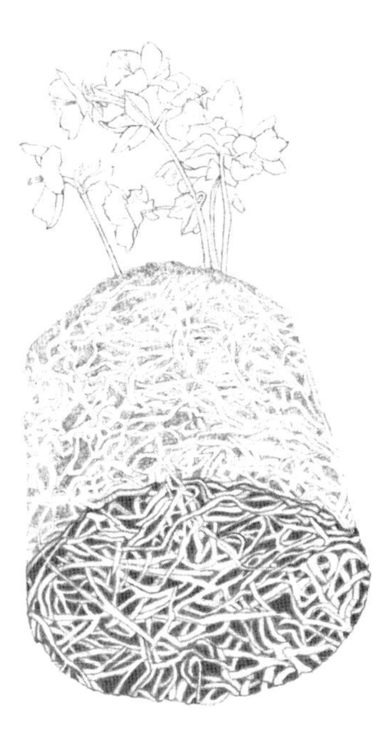

This plant is seriously root-bound. It looks creepy, but it can be salvaged. Cut away at least half of that snarl, loosen up the rest, and replant.

Filling the Containers

Now you have everything—potting soil and its additives; seeds and plants; containers and tools—all laid out and ready. Take a deep breath and dig in.

First, get the soil ready. In your large dishpan, dump in a big portion of potting soil, and add a small handful of water-holding gels and slow-release fertilizer beads. Slip on the thin plastic gloves, and mix that all up. Gradually add water, mixing as you go. To check that it's ready, take a handful of soil and squeeze it into a fat sausage shape; if it holds together after you open up your fingers, you're halfway there. Now poke that sausage with a finger and watch what happens. If it breaks apart, this mixture is ready. If it stays whole, or if water seeps out through the finger depressions, it's too wet and you need to mix in more soil. When the mixture is just right, add soil to your container up to about the three-quarter level. This will give you room to add your plants and their root balls, backfill around them, and still leave at least a one-inch open space at the top for watering.

Despite what you might have heard, do *not* add a layer of small rocks in the bottom. There's a longstanding myth that this helps drainage, but that is simply not true; in fact, it usually interferes with drainage, due to complexities of soil structure. If you are adding something heavy to your containers because your garden area is subject to high winds, use one or two large items like bricks and keep them away from the drain hole.

DON'T ROCK THAT POT

Time to retire this old wives' tale. Do not put rocks or broken pottery or gold nuggets or anything else in the bottom of the container. It does not "promote drainage"; in fact, just the opposite. *Drainage*—the downward movement of water—happens automatically; it's called gravity. Whether or not it drains *smoothly and freely* is a matter of the soil's structure, and has nothing to do with rocks on the bottom.

In one of those mysteries from the garden gods, soil does not really escape through the hole. But if you need a little peace of mind, you can lay a trimmed coffee filter over the hole before you start filling in the soil.

Now arrange your transplants, still in their nursery pots, on the top of the soil (or on the floor near you), until you get the positions just right, and start digging, one hole at a time. Remove the plants carefully from their nursery pots, keeping the root balls and soil intact as much as possible, unless you need to do surgery on a root-bound mess. Place the first plant down into the hole at the same depth it was in the nursery pot (tomatoes are an exception, as you'll see in chapter 5) and backfill with more handfuls of potting soil, pressing gently to hold the plant steady. Then move on to the next. When everything is in place, water gently to help the plants snuggle into position, and that's it. If your plan calls for seeds, too, put them in now, following the directions on the packet for depth and spacing.

Here's a technique I worked out some years ago especially for planting from six-packs. First I fill my former milk-jug watering bucket with water and make a very weak solution of fertilizer, maybe one-quarter strength. Then I remove the baby plants one at a time (that plastic is very thin and releases the plant if you just squeeze it) and soak each one in that water jug while I get its hole ready. That short dip in the water loosens up the root structure just enough. I get that first plant into the soil, squeeze out the second plant, soak it briefly while I dig the second hole, and so on. When all six are planted, I pour the contents of the jug around them to help them settle in.

Keeping Everything Handsome and Healthy

I wish I could tell you that because you did all that careful planning and precise planting, your garden is going to take care of itself from now on. But I know you know that's not true.

This is what gardeners mean by a *six-pack*. No beer in sight.

If you ever find yourself grumbling about the watering chore, think about what wartime gardeners had to do.

I grew up on a farm in mid-state Illinois, one of five children. We always had gardens on the farm, but during the war years they became Victory Gardens. In fact, we had three separate gardens: the truck garden (large), the smaller kitchen garden, and the kids' garden. In reality, all of us were involved to some extent in all the gardens, and sometimes the dividing line was a little fuzzy.

The "truck patch," as we called it, had a lot of sweet corn, potatoes, tomatoes, watermelon, and cantaloupe (which mostly never really ripened). Most of the produce from this garden was destined for storage for wintertime: the corn and tomatoes were canned, and the potatoes were stored in the cellar. I can clearly picture the rows and rows of colorful quart glass jars. The kitchen garden was closer to the house and had all the veggies you would expect in an Illinois garden: tomatoes, carrots, peas, radishes, lettuce, turnips, spinach, chard (which we mostly fed to the chickens).

When we kids decided to do a Victory Garden, we mostly chose tomatoes, radishes, carrots, and lettuce. It was close to the kitchen garden so that Dad could make sure we watered it well in the hot Illinois summers. We were supposed to take complete care of that garden, from planting to harvesting, but I can't honestly say we did.

What I mostly remember was the watering. It was especially important that the kitchen garden and truck patch were kept well watered, since they were major food sources. It was an arduous process, and in the dead of summer it took all of us to keep up with the blistering hot weather. Often we had to do it every day. We had a well with a long-handled pump somewhat near the kitchen garden, so everyone just grabbed whatever buckets we could carry and filled them with water. For the truck garden, Dad filled large containers, loaded them onto a wagon, and hitched up Buck and Nell, our draft horses, to move them to the big garden. Did I mention that summers get hot in Illinois?

Sandy Stonebreaker
Ocean Park, Washington

During World War II we were living in Claremore, Oklahoma, birthplace of Will Rogers, about thirty miles from Tulsa. My dad was a welding inspector with the Atomic Energy Commission, which meant he got sent to various locations all over the US. Sometimes we moved with him and sometimes we stayed behind in Oklahoma. All around us, everyone planted Victory Gardens, including my mom. But the real story was my Uncle Johnny's garden next door.

Uncle Johnny held a middle management position with an oil company; it was considered an essential job, and that made him ineligible for the draft. Uncle Johnny was a very fastidious man. His dress was always impeccable and his bow tie "just so." These same traits showed up in everything he did. His Victory Garden was something to behold. He spent hours digging and redigging the soil. Then he raked the garden bed in several directions until there were no clumps, just beautiful, loose soil that would flow through your fingers. Next he worked fertilizer into the soil. He chose chicken manure because he considered it the best, even though it was smelly. Again he worked the soil until it was beautiful and smooth. Finally he planted seeds. The sprouts sprang from the ground, alert and green—until they all curled up and died. He had overfertilized and burnt the plants.

Kay Ward
Portland, Oregon

I *can* tell you, however, that as a container gardener you are going to have a much easier time with all the ongoing chores than your friends with big in-the-ground gardens.

For one obvious difference, your garden is much smaller, and therefore less maintenance time is needed; that's simple arithmetic. But there's another, subtler difference. Because of its location on your balcony, deck, patio, or porch, your garden is closely linked to your living space. You move between the two frequently and seamlessly, and I'll bet you any amount of money that you cannot help but stop to pet the plants as you pass by. At the same time, almost without trying, you are also checking for early signs of trouble, when problems are simpler to deal with.

ONGOING MAINTENANCE

Watering. This is one critical distinction between container gardening and gardening in a traditional plot. In a big garden, each plant is surrounded by a large buffer of soil from which to draw moisture. Your containers are surrounded by air—no buffer at all, just empty space. The only moisture they will ever have is what falls from the sky and what you, their steward, add. This is going to be your biggest, most frequent chore, and without question the most important. It's not tricky; it just takes a little discipline.

When should you water? Not trying to be a smart aleck here, but the honest answer is *whenever it's needed.* You have a terrific measuring device right at hand—ten of them, in fact. Stick a finger into the soil of a container as far as you can. If the soil is dry, it's time to water. Another reliable clue: the plants themselves will tell you. If they start to droop, they *really* need water, so get busy.

This is going to vary, of course, depending on the season and the weather on any given day. In the middle of summer, when the temperatures are very high and there has been no rain for a while, you may need to water every day.

You will never eliminate the need for watering, but there are several ways you can make things a little easier on yourself. One inexpensive solution is those magic water-holding crystals you met in chapter 3; they really work. You might want to look into what are described as *self-watering pots*; they rely on a built-in water reservoir at the bottom and some device to wick the water upward. Another possibility, more of an investment, is a drip watering system, composed of a slender hose with offshoot emitters that clip individually onto pots and are activated by a timer attached to the water spigot.

A very effective approach, and one that costs nothing extra, is to use the largest containers you can comfortably manage. By definition they will hold a larger amount of potting soil, and that means that each individual plant has more buffer against drying out than a similar plant in a small container.

Fertilizing. We covered the fertilizer basics in chapter 3; our focus now is how to apply them and when. Whatever product you purchased, its package has instructions for how to prepare and apply. Do that. Do not succumb to the temptation to mix a stronger solution; you'll just end up burning your plants.

But if those instructions also tell you how often to add the fertilizer, you have my permission to adjust that. Here's why: Those package instructions are meant to serve all gardeners, but container gardeners have a special challenge. All fertilizers, no matter what physical form they started as, need to be dissolved in water in your containers; that's the only way the plant roots can pick them up. But with all the watering you're going to be doing in hot weather, you'll also be washing away fertilizer. So it needs to be replaced. Another consideration: plants need a fertilizer boost when they are in active production mode, which for many of our most popular vegetables coincides with the warmest weather, which is when they most need watering and, therefore, fertilizer replacement.

Here's a system I worked out for myself a few years back. When the weather is really hot and plants need to be watered frequently, I mix up a very weak solution of fertilizer, perhaps one-quarter strength, and set it aside. Then I alternate watering with that mixture and plain water, like this: plain water, plain water, fertilizer water. In my climate, that cycle usually carries me through one week in summer.

Weeding. We can dispense with this one very, very quickly. You aren't going to have many weeds. For one thing, you started with clean, weed-free potting soil. For another, you paid attention when I told you to keep the containers packed with plants, so even if a few weed seeds do manage to sneak in (thank you, Mr. Robin), they won't have room to get established. For still another, you're going to be admiring your containers close up all the time, and you will quickly spot any gate-crashers that think they can get by you.

POSSIBLE PROBLEMS: PESTS AND DISEASES

Here, too, the fact that your garden is relatively small gives you a great advantage. Because everything is so close at hand (not way out in the back forty), any problems are easy to spot and thus quickly dealt with. Let's run down the most common, in two groupings: harmful insects and plant diseases.

Damaging pests. The two you are most likely to encounter are slugs and aphids. There are other bad boys—cutworms that slice the whole plant right off at the soil line or hornworms that feast on your tomatoes—but in my experience they don't easily find a foothold in container gardens. But if your climate supports slugs, you could well find some lurking around. And aphids, I do believe, could find plants on the moon.

Slugs. Imagine a snail without shell; that's what slugs look like. They thrive in cool, damp environments, so you may be more likely to see them in the spring. Except you may not see them at all; they come out to party at night, chomping on the foliage of just about anything in their path.

Your first clue is big holes on the leaves of your plants; it's unattractive, but it's also a serious warning. Several slugs attacking one plant for several days can so weaken that plant that it simply collapses.

Here are two simple, nontoxic fixes. Position some solid item flat on the soil near the holey plants; a piece of heavy cardboard, for example. The next morning you will find a cluster of slugs on the underneath side; toss the whole thing into the garbage. The other one leads to many jokes but works quite well: in the evening, sink something like an empty tuna-fish can into your container down to soil level and fill it with beer. The next morning, you'll find dead bodies floating. (According to the jokesters, they died happy.)

Aphids. These tiny little bugs love to feast on the tender new growth of your plants so much they clamber over one another to get their share. A heavy infestation will kill the plant. They are usually green and tend to cluster on the undersides of leaves, so their presence is not always obvious. A good clue is a recurring line of ants into your container: the aphids secrete a sweet goo that draws the ants and coincidentally alerts you to the problem.

Because a big cluster of these things is so creepy looking, your first impulse may be to look for some kind of insecticide. There's an easier solution: Slip on one of your disposable plastic gloves and simply wipe the aphids off, crushing as you go. Turn the glove inside out to seal the dead bodies inside, toss it, then wash that leaf with plain water. Even simpler: cut the affected leaf off entirely and put it straight into the garbage. Watch for a return of the ants, check other leaves now and then, and otherwise don't worry.

Two common diseases. The full list of plant diseases is long, but in reality there are just two that you are likely to encounter—mildew and blossom end rot. It happens to the best of gardeners, so keep your eyes open.

Mildew. You've probably dealt with mildew at some point in your life, and so the plant version is pretty easy to recognize. There are two forms: downy mildew, which shows up in cool, damp environments and mostly affects cool-season vegetables; and powdery mildew, a summertime problem, most often found with squashes and cucumbers.

If left unattended, the infected plants will eventually collapse. So first, cut off and dispose of the leaves with obvious infection. Then protect the rest of the plant by spraying it with a solution of one teaspoon of baking soda dissolved in one quart of water; repeat that spray a few days later, just to be on the safe side.

Blossom end rot. This one is also easy to recognize because the name is so descriptive. At the end of the fruits where the flowers used to be, a big, ugly black spot develops. This most often shows up with tomatoes, to a gardener's heartbreak, and can be frustrating to control. Two common causes are uneven watering and calcium deficiency in the soil; in both cases, by the time you see the signs, the damage has been done. Your best bet at that point is to toss the ruined fruits, correct either shortcoming, and cross your fingers.

A little perspective. It's not fun to think about the plants you have so carefully nurtured suddenly becoming infected with a nasty disease or being attacked by mean little critters. Of course you want to step in and fix things. And if that first attempt doesn't work, you probably wonder whether there's something else you should do, something stronger you could try.

But let's pause for a moment and consider. It's just a plant. You did the best you could without adding any toxic substances. Good for you. Give the corpse a decent burial, and then go to the nursery and treat yourself to something new.

CHAPTER 5

The Good Stuff: Vegetables

This chapter, and the next two, are the heart of the book. Here is where you will find information about the plants you're thinking of growing, ideas for specific varieties, and advice on how to be successful with them.

I'm sure you will want to know what the individual varieties look like, especially if I'm recommending something that is new to you. In the back of the book, you'll find a gallery of color photographs of many plants, organized in the same sequence as the plants appear in this chapter and the next two. Not everything, of course, is included—only the ones I think you may not be familiar with. You already know what a radish looks like, but how about a watermelon radish? Or one with a bright magenta coloring? Reading my description is just not the same as seeing a beautiful picture, and that's the purpose of this photo gallery.

First, though, this may be a good spot to summarize a few broad guidelines:

- If this will be your first year as a container gardener, I suggest that you choose transplants rather than seeds wherever possible. This might mean that you have to pass up brand-new or unusual items that are not yet available in your local garden centers; it will also mean fewer frustrations. Next year, with experience under your belt, you can expand your horizons and shop the seed catalogs to your heart's content.
- Remember that when I name a specific source for suggested varieties, it's because, as far as I now know, they are not available elsewhere. If no source is mentioned, it means seeds (and sometimes mail-order plants) are available from multiple sources.

They might also be available as transplants at local nurseries, so keep your wish list handy when you go.

- Remember that you have one warm season (summer) and *two* cool seasons (spring and fall). That probably seems simplistic, but it is surprisingly easy to lose track of how your garden plans match up with the calendar. Also, remember that some things can span across seasons, if the garden gods are smiling.
- Remember, too, what this is all about, and keep your perspective well polished. Some things will be wildly successful; some won't. Relish the successes, make notes about any disappointments for next year, then don't fret.

ASIAN GREENS

One of the greatest delights of growing your own vegetables is being able to have right at hand those special items you might not easily find at your regular supermarket. In that category I happily put a family of vegetables I have grouped together as Asian greens. If you're not familiar with these wonderful plants, I hope you will try one or more. They are easy to grow, very nutritious, adaptable to many dishes, and delicious. A satisfying plus for container gardeners is that many of them have a compact, vertical growth habit that means they can be planted close together and even when fully grown will take up very little space in comparison to the edible parts—which is the whole plant. Ergo, ratio winners.

A note about organization: because these plants are so similar, I decided to present them as a group, both to save space and to avoid annoying you with repetition. There is a separate listing for leafy greens, the ones we usually eat in cooked form—spinach, kale, Swiss chard, and mustards. Lettuce, together with the other greens you might use raw in a salad, also has its own section. It's all a bit arbitrary, I realize; if you're looking for something specific and don't find it where you think it should be, check the index.

Asian Greens Basics

- These are cool-season plants that will often carry through from spring to early summer. For extra insurance, look for varieties described as "slow to bolt" or "heat tolerant."
- Start with transplants as early in spring as you can find them.
- In late summer, start a fall crop with new transplants, if you can find them, or seeds if you cannot.

- They prefer full sun but will also do quite well in partial shade.
- They are happiest with steady watering and a taste of nitrogen-rich fertilizer every now and then.

GROWING TIPS

These plants are both sturdy and tender. They're sturdy because of overall vigor. They can handle a wide range of temperatures except very hot, aren't much bothered by disease, and don't need staking or any other kind of physical support. I call them tender because that's how I think of their taste. The stems are crisp and juicy; the leaves are tender and spicy-sweet. You feel like you are eating pure vitamins—because you are.

To achieve that crisp/sweet continuum, the plants need to grow quickly and steadily. That means scrupulous attention to watering and a regular dose of nitrogen-rich fertilizer. Nitrogen, you will recall, stimulates leaf growth, and these are leaf-total vegetables.

All these greens are ideal for starting a second crop in the fall. You may not find transplants then as easily as in spring, so this is a perfect time to practice your seed-starting skills.

For seeds, you have a couple of choices. You could clear off a space in one of your containers, designate a small container just for growing seeds, or set up a peat pellet tray (see chapter 4) in some sunny spot. (I vote for the peat pellets.) Then, by the time the seedlings are about three inches high, you should have some empty space in your big container, because the summertime plants have reached their natural end.

HARVESTING

To ensure you enjoy the full richness of crisp/tender/sweet, it's important to harvest these vegetables before the stems get tough and stringy and the leaves get leathery. It's not difficult; the hardest part is remembering to pay attention to these fast-growing beauties. You can expect to start harvesting in as little as three weeks after planting.

If you are growing the full-size plants, rather than the "baby" varieties (see following pages), and if this is your primary vegetable for tonight's dinner, you might want one whole plant, maybe even two depending on how many people you are feeding. Dig the plants up carefully or slice off the entire plant at the soil line.

Another option is to remove only as many outside leaves as you need for whatever you're cooking tonight. Then the next layer of leaves, which are now the outermost, will be ready for harvest within a few days, freeing the next inner layer to develop. Remember

that time you bought Chinese cabbage at the supermarket and it turned into black slime before you could use it all? With these greens in your garden, harvested in this few-at-a-time process, you'll never have that again.

To remove the leaves, either grasp the stem at the lowest point and gently pull outward and down (not up), or slice them off at the base with a sharp knife. With the second technique, you're less likely to accidentally pull the whole thing out of the ground.

You can also harvest some of those full-size plants early, when they reach five or six inches in height, and you will have in effect baby versions of whatever it is. At this young age, they are wonderfully tender and sweet. You can either dig up the entire plant and slice away the roots, or cut the plant off about an inch above the soil line. With the latter, there is a good chance you will see new growth forming from the old base.

In recent years, several plant breeders have developed baby versions of some of these plants that are meant to stay small even at mature size. The same growing conditions apply: steady watering, a little nitrogen fertilizer, harvest before they get tough. You will probably want to harvest the entire plant at one time, since they are so small and cute.

Varieties

CHINESE CABBAGE

In one sense, all these Asian greens qualify as Chinese cabbage, since they originated in Asian countries and they are all members of the cabbage family, *Brassica*. And because their common names are English transliterations of Asian words, you will encounter various spellings. Don't let that interfere with your enjoyment.

The vegetable known most commonly in the US as Chinese cabbage is a tight bundle of pale green leaves, taller than wide, with a taste somewhere between celery and lettuce and cabbage. (It is also called *Napa cabbage*; don't ask me why.) If you ever feel inspired to try making kimchi at home, this is the cabbage you want.

CUT AND COME AGAIN

Many of the vegetables that we grow primarily for their leaves respond well to a harvesting process called *cut and come again*. It's just what it sounds like: cut off some of the leaves, and others regrow in the same spot. If you've never tried it, be prepared to be amazed.

It's unbelievably simple. Reach down to the plant in question and grasp a handful of leaves. With a sharp knife or pruners in the other hand, cut away that handful about one inch above the soil line. If you need more for tonight's meal, move to another side and harvest another handful. The plant will grow new leaves in those spots. Of course it doesn't happen overnight, so you want to stagger your cuttings in various spots so that no one area looks mohawked.

Some well-established varieties, widely available as seeds, are **Blues**, **China Pride**, and **China Express**. **Little Jade** from Renee's Garden Seeds is a baby version, less than half size. Another good mini is **Minuet** from Johnny's. For the pure fun of it, think of adding red-leaf varieties in with the more traditional green. Some beautiful possibilities are **Merlot** and **Scarlette**.

MIZUNA

We might think of mizuna as a crossover: it can be enjoyed both raw and cooked. The distinctive leaf shape and slightly peppery tang earn it a spot in your fanciest mixed salads, and contribute a pleasant-tasting green veggie to a stir-fry. I grow it because it's so darn pretty. Given enough space, the plant grows into a rosette shape, which is most apparent when viewed from above, so I like to plant it in a lower container. In recent years a new version has appeared, with brilliant magenta stems; a rosette of *that* would be stunning. Among the new varieties, all quite similar, are **Red Streak**, **Central Red**, and **Red Kingdom**.

There's more. Mizuna outshines most of its cousins in its ability to survive warm weather. Plus it reaches harvest size in a little more than a month, so you can start it in early spring, replenish several times with new transplants, and enjoy this little beauty through everything but the worst heat of summer, then start over again in late summer for a fall crop. One more thing: this plant lends itself easily to the cut-and-come-again style of harvesting—cut off a handful to about an inch above the soil line and watch new growth sprout from the cut areas. Garden magic.

In the garden centers, mizuna transplants are often stocked with the lettuces. As is often the case, they are more difficult to find for the fall planting season. But mizuna is not at all tricky to grow from seeds, which are readily available.

PAK CHOI/BOK CHOY

Regardless of how you spell it, it's almost two plants in one. The snow-white stems, broad at the bottom and narrowing upward toward the green leaves, are crunchy and sweet, with a crisp texture similar to radishes. With that broad base, you could use the stems on your next tray of raw veggies for dips; see who can guess what it is. The rich, green leaves, a little bit spicy and a little bit sweet, are a fine addition to soups, stir-fries, or any other Asian dish you have up your sleeve.

Pak choi is widely popular, with good reason, which means it is also widely available as both seeds and transplants. Look for the variety named **Bopak**; it's an All-America Selection, so you can be confident it will do well in your garden.

You can now also find several baby pak choi varieties, bred especially to stay small. Look for **Hedou Tiny** and **Baby Milk** from Baker Creek, **Toy Choi** from Park Seed, and **Green Fortune** from Renee's Garden. By the time you're reading this, there could well be many others.

TATSOI

Tatsoi is actually a variety of pak choi, in botanical terms, but it is so different in appearance we often think of it as a separate species. The smallish, spoon-shaped leaves grow in concentric circles from a central core, forming the most gorgeous rosette. It's so dang lovely you may be reluctant to touch it. Harvest leaves from the outermost circles and the plant will continue to grow. Use it as you would pak choi; the taste is similar but stronger, most people think.

BEANS

You might call them *string beans*, although the "strings" have been bred out of most varieties. Or *snap beans*, because that *snap* is how you know they are ready to harvest. Or *green beans*, although you can also grow green beans that are purple or red or yellow. Let's just call them wonderful, easy, and delicious. And they are especially delicious from your garden because fresh beans lose their quality very soon after picking, and so only people who are clever enough to grow their own ever experience the truest flavor.

Important: within this large category are two big groups: pole beans and bush beans. Pole beans grow as long vines that need the support of something tall they can wrap around; bush beans grow as short, stocky plants that don't need that support. Container gardeners, stick with bush varieties. You're going to be buying seeds, and you'll have lots to choose from; just make sure they're all the bush types.

Bean Basics

- Give them full sun and regular watering.
- Start with seeds. You might find transplants for sale occasionally, but they don't work nearly as well as you'd like them to. And the seeds are so easy to handle and so quick to germinate, there's no reason to hesitate.
- Double-check the seed packet: Is this a bush type?
- Plant seeds one inch deep and three inches apart, starting in late spring (about two weeks after your spring frost date).
- The first harvest is ready in fifty to sixty days.

GROWING TIPS

When I said beans were easy, I wasn't kidding. Most modern varieties are disease resistant, and garden insects usually don't find them appealing. Your biggest concern will be watering. Once these plants kick into full production mode, they need a steady regimen. Be careful to water the soil, not the leaves; wet leaves can trigger a bean fungal disease. For the same reason, don't work around the plants or harvest beans right after rainfall.

Most varieties will produce all their beans over a period of about three to four weeks. To keep beans coming for a longer harvest time, some gardeners sow seeds in a succession, a week or so apart. You'll probably want to retire the plants that have reached their natural end, to free up space for the newcomers.

HARVESTING

Let's talk about when, and why, and how. The traditional advice about harvesting is that the beans are ready when they snap in half if you bend them (hence the name). That advice still works for the classic varieties, but some of the newer slim filet varieties are meant to be enjoyed when pencil-thin. If you're uncertain, harvest a few and steam them quickly; your taste buds will give you the answer.

The "when" problem you may have is doing it often enough. Beans move so quickly through their cycle—from pretty little flower to faded flower to teeny bean pod to the full-blown thing ready to be picked—it will take your breath away. Once the plant hits that full-on production mode, it will pump out new harvest-ready beans every day, like an automatic, unstoppable machine. If you don't keep the beans picked, that marvelous machine grinds to a halt.

And that's the "why." Think back to when you planted these beauties; the seeds looked like beans, didn't they? That's because they are. The seeds are beans; the beans are seeds. The bush bean plant in your garden desperately wants to keep producing pods with the little seed beans inside because, like all plants, somewhere deep in its genetic memory it knows it needs the seeds for survival of the species. If you interrupt that cycle by picking the bean, the plant will just make more and then more—which is, of course, exactly what we want. But if you happen to leave the beans on the plant so long that they mature to viable seeds for the next generation, the plant says to itself, *Ah, job done. I can quit now.* And no more beans for your dinner table. To make matters worse, the beans have a way of hiding under the leaves, so even if you're being careful, they can get overdone before you even spot them.

The "how" might seem counterintuitive at first. Bean plants look full and sturdy but have very shallow roots. If your instinct is to grab onto a bean and yank, I guarantee you will do it only once; there's a very good chance the entire plant will come out of the soil. Instead, use both hands. With one hand, hold on to the plant's main stem; with the other, gently pull a bean away with an upward motion. Some people find it easier to use their hand pruners instead.

Varieties

Your grandparents grew green beans long before you were born. Those same varieties are still available, plus today we also have some marvelous new items to complement the longtime favorites. Two well-established classics are **Blue Lake** and **Kentucky Wonder**. Recommendations from always trustworthy All-America Selections: **Topcrop**, **Greencrop**, and **Derby**.

A newish variety, still in the category of "green beans that are green," is the French filet bean, also known by its French name, *haricots verts*. They are ready to be harvested when still very thin—no more than a quarter-inch in diameter, about the size of a pencil. Several companies offer seeds, usually with generic names like haricots verts or French filet. One that stands out, in my view, is **French Mascotte**, from Renee's Garden Seeds. The beans themselves are the typical filet style, pencil-thin and incredibly tender; what makes these worth your attention is the way they grow. The flowers, and thence the beans, sit on top of the foliage cluster, which makes the beans not only easy to pick but also extremely easy to keep an eye on. Which is, you will recall, the biggest problem with standard bush varieties: you need to keep harvesting the beans, but they are easy to overlook because they are covered by the foliage.

And don't forget that not all green beans are green. Once you start exploring you will find varieties with red or purple pods, which are fun because they're unusual and also very attractive. **Red Swan** is available from several seed houses; for purples, look for **Royal Burgundy**, **Purple Queen**, or **Purple Teepee**. I'm intrigued by **Dwarf Velour French** from Park Seed; it's a French filet–type with a soft purple color. Now the bad news: lovely as they are, all these colors turn green when cooked.

At age ninety, Dobby grew enough marionberries to make 264 pies. In case you're not familiar, marionberries are a special type of blackberry, bred by the plant geniuses at Oregon State University. They are unique to the Pacific Northwest and they don't travel well, so only those of us smart enough to live here ever get to enjoy their amazing goodness.

I was born in my family's home in Nahcotta, Washington, in March 1931. We always had a vegetable garden, about half an acre, along with one or two milk cows, a pig, chickens, dogs and cats, ducks, and usually one or two beef cattle. In the family garden we grew potatoes, corn, beans, peas, lettuce, cabbage, carrots, berries, and flowers. We also grew our own hay for the cattle. Everyone in the family—Father, Mother, my two older sisters, and me—pitched in with everything.

After World War II began, families were encouraged to grow Victory Gardens to help with the war effort. Our family had always been involved with 4-H, and I perked up when I heard that 4-H was holding a contest for the best Victory Garden. I really wanted to win a prize, so I spent a lot of time working my share of the family garden. The county extension agent, Ralph Roffler, came to judge my garden. He checked everything carefully, and under a bean plant he found one weed. I was certain that eliminated me, and I was crushed. But everything else must have met his standards, because I won a blue ribbon.

As a result of my Victory Garden experiences, I have always grown a garden when possible, and I'm still doing it at ninety years old. Last summer [2020] I picked enough marionberries for 264 pies and about 1,500 zucchini from fifty plants.

Dobby Wiegardt
Nahcotta, Washington

BEETS

I suspect that beets fall into that vegetable category known as "you either love them or hate them." You know which group you're in, and that's OK with me; no judgments here. (I will confess that there's an item in this chapter I honestly can't stand, but I'll happily tell you everything you need to know about growing it. I just won't tell you which one it is.)

On the other hand, I happen to love beets. I love their tender, fresh-tasting greens; I love their gorgeous color; I love their unusual varieties, so surprising and fun that even beet-haters are tempted to try. I even love their funny-looking seeds.

Beet Basics

- Grow in full sun or partial shade.
- These are cool-season plants.
- Start with seeds directly in the container soil.
- Baby seedlings need thinning. Don't argue.

PLANTING AND GROWING TIPS

Let's talk about those seeds. They are on the small side, about the size of a shriveled-up pea, and very hard and bumpy looking. Turns out, what appears to be one rather odd-looking seed is actually a cluster of several seeds organically fused together. So when you plant one seed, you're going to get several plants, maybe as many as five. You *cannot* let all five stay in place, or none will grow satisfactorily.

Plant the seeds about two inches apart. Then when the little seedlings in one cluster have grown to two to three inches in height, choose the strongest-looking one and remove all the others. This is going to be hard; do it anyway. You can then, if so inclined, add them to tonight's tossed salad. If anyone asks, you smile and say, "Oh, I grow microgreens." Because, well, that's exactly what you did.

The seeds will germinate more quickly if you soak them in plain water overnight; that will soften the hard outer coating. Beets will typically be ready for harvest in fifty to sixty days. If you are very fond of them or interested in trying several varieties, you may want to stagger planting at several intervals, starting in early spring, for a longer harvest period. Then in late summer, do one last planting for a fall and winter crop.

The secret to beets with a pleasing texture is consistent watering and timely harvesting; to keep them from becoming tough and woody, do not let the soil go dry and do not let them grow overlarge—simple as that.

HARVESTING

You can harvest the leaves a little bit at a time for a colorful addition to tonight's salad; the plant will produce more leaves. For the beets themselves, the best technique is to gently brush away the soil from the very top of the beet, where the leaves emerge, until you can see the entire circumference. Does it seem a reasonable size? If so, tug it up. If not, pat the soil back in place and wait a few days.

Here's another approach, warmly recommended: harvest some when they are still small, say one inch in diameter. At that size they will be unbelievably tender and sweet. Leave some of the others from the same planting to reach their mature size.

Varieties

These two-for-the-price-of-one plants deserve a place in your garden for all the reasons I noted above, and here's one more: they take up so little space that you can add some of the unusual, fascinating varieties just to trick someone who thinks they hate beets. Or because they are not only delicious but beautiful.

Especially good choices for container gardeners are varieties bred specifically for their small size at maturity—look for **Baby Ball** and **Babybeat**. For the classic red round that you're already familiar with, try **Detroit Dark Red**, an heirloom variety first introduced in 1892; seeds are widely available. Other possibilities for the standard shape and red color:

Sweet Merlin, bred especially for roasting, with its high sugar content; and **Red Baron**, from Renee's Garden.

You can also find beets that are cylindrical in shape, rather than small, round balls; one good variety is **Formanova**. The relatively slender, longer shape allows you to slice these beets horizontally into coins of all the same size.

Still in the "red" category, but in a class by itself, is **Chioggia**, an Italian heirloom known to European growers since 1840. The flavor is mild, mellow, and sweet—completely unlike the earthy taste that many people associate with beets. But the real glory is the coloration. On the outside, it looks like an ordinary red beet. But once you cook it and slice it horizontally into rounds, you have revealed a stunning, altogether unique pattern: concentric circles of white and candy-cane red, pretty as can be. Present this to your beet-scorning friends, and watch for their reactions.

That's not all. Several varieties of beets with golden-yellow flesh are available. One, called just **Golden**, dates to the 1820s. Also **Touchstone Gold**, available from several sources. One very nice feature of the golden varieties is that they do not "bleed" when cooking. (You can minimize that unpleasant trait with red varieties if you leave an inch or so of the leaf stems in place during cooking.) And just to keep things interesting, you can also find varieties with white flesh. Look for **Albino** or **Avalanche**. Try to fool your friends with that one!

You can create a lovely presentation with a platter of red, yellow, white, and Chioggia beets all served together. Several seed companies will give you a head start; they have created a beet collection with different types of seeds all together in one seed packet.

CARROTS

You have some in your refrigerator right now. About eight to ten inches long, bright orange, probably been there for a while. But if you want to know what a fresh carrot really tastes like, you have to grow your own.

Fortunately, we container gardeners have several delightful possibilities to choose from, all of them items you will never, ever find in the supermarket. Unfortunately, the terminology that horticulturists use to describe types of carrots is confusing, so this is one time I'm hesitant to suggest you get your bearings by spending time with a few seed catalogs. Instead, I'm going to point you toward a more casual descriptive language and then to specific varieties that fit within those groups.

Carrot Basics

- Likes full sun.
- Grows best in cool seasons: spring to early summer, then again in the fall.
- Start with seeds, and direct-sow into container.
- Slow to germinate, so learn a few tricks.
- Needs consistent watering.
- Very few diseases.

PLANTING TIPS

The hardest part about growing carrots—indeed, the only hard part—is that the tiny seeds are very slow to germinate: ten to twenty days, on average. In the meantime, their soil bed must be kept evenly moist or else nothing happens; you'll be patiently waiting in vain. Here are a few tricks.

- In a small bowl, mix some radish seeds in with your carrot seeds. Approximately half and half, although precision is not important and, in any case, the carrot seeds are so tiny you couldn't count them if you wanted to. Mix the two together very well, and plant the mixture about a half-inch deep. The radishes will sprout much more quickly, showing you exactly where the carrot seeds are and keeping the soil loose. Keep up with the watering; the baby radishes need it to stay crisp and tender, and the carrot seeds *really* need it.
- You can also plant carrots by themselves (about a quarter-inch deep) but add something on top of the soil to keep the surface moist. Something like a strip of burlap, predampened peat moss, even a layer of paper towels or thin cardboard. Every day, lift up your covering and check the soil. If the surface is not damp, add water (use a spray bottle so the seeds don't get dislodged), replace the cover, then spray the covering as well. Every day.
- When the little green darlings start to show themselves but there are just a few and then no more for several days, add some more seeds. Ultimately you may end up having to do some thinning, but the new batch will catch up quickly.

GROWING AND HARVESTING

Once you have the tiny seedlings on their way, your work is practically done. Not kidding; everything else is easy. Carrots are not prone to major diseases, and harmful insects don't seem to like them. The only thing you, their guardian, have to focus on is watering. Like most root vegetables, carrots depend on a steady, consistent supply of water for a tender texture.

Harvesting is also not complicated. If you know the anticipated harvest time ("days to maturity" on the seed packet), start checking by easing one out of the soil. Grasp the plant at the base of the foliage, give it a little wiggle back and forth, and pull gently. To enjoy the full flavor of these beauties, you'll want to harvest carrots the same day you plan to serve them. But if plans go awry, the carrots will keep in your refrigerator for a couple of weeks. Here's another approach, well worth trying: harvest carrots *before* the official maturity time frame, and you'll have exceptionally tender and sweet baby carrots.

Varieties

The language of carrots is confusing. New varieties are described as "Chantenay type" or "Nantes type," which is distinctly unhelpful if you don't know what Chantenays and Nantes themselves are like. I'm going to use a simpler nomenclature, built around categories of carrots that work especially well in containers.

ROUNDS

Rounds are exactly what you would think from the name: little round balls of orange sweetness. They are also incredibly cute, guaranteed to delight the children in your life. When my nephew was about four, he announced that he was done with vegetables. Nothing would change his mind. At one point I asked him, "Honey, do you like *anything* green?" He thought for a moment and then answered with a big grin: "Apples!" I wish I had known about these carrots in those days.

Here are three to consider: **Thumbelina**, a widely available All-America Selection; **Parisienne**, a French heirloom from several sources (**Parisian** is the same); and **Romeo**, from Renee's Garden. Harvest all these when they are about a three-quarter-inch diameter. This usually takes between fifty-five and sixty-five days.

FINGERS

Another wonderfully descriptive term. They are slender and long rather than short and round. But "long" is relative; these are bred to be perfectly ready when they are about as long as your finger. The best known is **Little Fingers**, widely available. Harvest when they're three to four inches long, which takes about fifty-five days. Another nice possibility, which fits loosely into this category, is **New Kuroda** from Baker Creek. This improved heirloom from Japan is finger-length but shaped like a short cone, with a wider top tapering to a narrow bottom. Especially noteworthy for its hardiness, it does well in all seasons, even hot summers. If you live in the South, take a look at this one.

LEAFY GREENS

In this section I have grouped together four types of plants that are similar both in the garden and in your kitchen: chard, kale, mustard, and spinach. I did this both to save space and to spare you the annoyance of reading the same things over and over. But I freely admit this grouping is arbitrary and unscientific; I was more concerned with how you might use these plants than how a botanist might characterize them.

Remember in our earlier discussion of ways to make the most of your limited space, I suggested incorporating the concept of ratio—how much actual foodstuff is produced by any one plant in relation to the amount of garden space that plant takes up while it's growing. Well, the plants described in this section win the ratio gold star, because we eat the whole thing!

These plants are also wonderfully easy to grow, with very few disease or insect problems to worry about. They accommodate a wide range of temperatures, from warm to really cold; many fans believe these greens actually taste sweeter when kissed by frost. They are handsome as well as tasty. And they are powerhouses of nutrition, with several vitamins, beta-carotene, even calcium. If some of these are new to you, I'm happy to introduce you.

Leafy Greens Basics

- These greens are happiest in full sun but won't complain about partial shade.
- As a group they do best in cool seasons, but many will tolerate the full range of weather patterns, from warm summers to very cold winters.
- Start with transplants if you can; direct-sow seeds if you cannot.
- Most are rapid growers, some as early as twenty days, although thirty to forty is more typical.

- Steady, consistent watering and a dash of nitrogen-rich fertilizer will keep the plants producing tender, succulent leaves.

PLANTING AND GROWING TIPS

Mature plants can reach a robust size, more than a foot in height and width. With that in mind, allow room to spread if you are planting transplants, or be prepared for thinning if you direct-seed. In either case, do your first planting in early spring and again in late summer for a fall crop.

The plants themselves are quick to grow to size, but you can help them do even better with steady watering and light applications of a fertilizer high in nitrogen. Pay attention: that fast growth means it's all too easy for leaves to get oversized and then tough. In the damp days of early spring you may have a few slugs, but in general these are problem-free plants.

HARVESTING

All these plants lend themselves very well to early harvesting, while they are still baby size, for a delicious, tender treat. Use a sharp knife to slice off the entire plant at the soil line. Leave others to grow to size; remove mature leaves one at a time, with a sharp knife at the base. All of these plants develop a stronger, more pronounced taste as they get larger. The trick is to not let them get too large, to the point they become tough. Once they are four or five inches tall, they are also very successful in the cut-and-come-again mode (see page 94).

Varieties

CHARD

I grew up in the South, where the term *greens*, as in "a mess of greens," usually means collard greens. My South Carolina relatives may disown me when I say this, but I cannot stand collards. So it was something of a leap of faith, years ago, when I agreed to give chard a try. Oh my heavens, am I glad. Easy to grow. Doesn't poop out in summer weather. Nutritious as all get-out. And so handsome you could grow it just for looks. The only difficulty is trying to make sense of the confusing names.

One very common question: What is the difference between chard and Swiss chard?

Short answer: for all practical purposes, there is none. The original plant is a Mediterranean native, known since Roman times; like many plants, it evolved in several directions over time. In the nineteenth century, so the story goes, a botanist in Switzerland identified two separate types, one of which came to be known as Swiss chard in his honor. In the intervening years, that type has so dominated the marketplace that nowadays chard *is* Swiss chard; the terms are widely used interchangeably. You will find Swiss chard as seeds from many suppliers, as baby transplants in most nurseries, and in the produce section of many supermarkets as individual leaves cut and banded together like a floral bouquet.

As for specific varieties, let's start with **Baby Leaf Chard** from Renee's Garden Seeds. Direct-sow in early spring, preferably in full sun; it'll be ready for harvest in thirty days. Use the cut-and-come-again method to keep the plants producing, or start a new batch of seeds, or both. It grows through all but the hottest summers, and its mild flavor makes this a nice green for salads when your lettuces are threatening to bolt.

The focus is all on the tender, sweet-tasting leaves of **Verde de Taglio**, an Italian heirloom from Baker Creek. It's another excellent producer that loves the cut-and-come-again style of harvesting.

A Swiss chard variety from Renee's Garden developed especially for containers, with green foliage and brilliant golden stems, **Pot of Gold** needs fifty days. Like other fast-growing plants, these will thank you for steady watering and regular fertilizing.

The magnificent **Bright Lights** needs fifty-six days. This is the one I fell in love with, the one that convinced me I was wrong about greens. Apparently I'm not alone in this; Bright Lights is widely and wildly popular, for good reason. In comparison to most other chards, the texture is tenderer and the flavor milder and sweeter. And oh my gosh, are they beautiful, with stems in a gorgeous palette of gold, pink, orange, purple, red, and white, with bright, pastel, and multicolored variations of all those colors.

Transplants begin to show up in your local garden centers in early spring, often in six-packs or four-packs; this is your easiest course. But if circumstances (or personal preferences) take you in the direction of seeds instead, you'll find them easy to manage. Plant directly in your containers one-half-inch deep and about one inch apart; they will germinate in about a week and be ready for harvest in about fifty days.

In the meantime, take inventory when the seedlings are about two inches tall and remove any that are too close to others whose color you like better. (Thinning is always good.) You can begin harvesting for baby chard when leaves are about four inches tall;

slice the stems cleanly with a sharp knife about an inch aboveground, and new leaves will regrow. At mature size, the plants can be as much as two feet tall, which is why your earlier thinning was important.

Among its other good qualities, Bright Lights does well in almost every season and climate, is very slow to bolt in hot weather, and doesn't wimp out in the cold. No wonder it was named an All-America Selection.

When you go shopping for Bright Lights transplants, you may find items with similar names: **Rainbow Chard** or **Five Color Chard** are common examples. I don't want to discourage you from them, especially if that's all you can find, but just to make you aware that those are *not* the same as Bright Lights. It's a bit like Dijon mustard: in order to be called by that name, the product has to meet certain precise specifications, and only Bright Lights is Bright Lights.

The Bright Lights information page from Johnny's Selected Seeds starts with this line: "The gold standard for multicolored Swiss chard." They have every right to brag: more than thirty years ago Johnny's began breeding this brilliant mix, originally created by an amateur gardener in New Zealand, and continues as its guardian, selecting and maintaining the various colors and producing the seeds.

KALE

Have you noticed? Kale is definitely having a moment. It seems to pop up in popular magazines and garden blogs all the time, and it's easy to understand why. It's very handsome, easy on the gardener, and definitely a vitamin powerhouse.

This new attention to an old vegetable is partly a result of our national focus on healthy eating, but also, I suspect, because of newer varieties in strong colors, which allow gardeners to plan a pretty tableau of colors, shapes, and textures. Check out some of these described here, and imagine what you could do with a kale-only container.

You might already be familiar with **Tuscan**. It's a classic, well-known and widely available in the supermarket as well as the garden center, and has an unmistakable look: long, narrow leaves whose edges curl inward, making the entire leaf look a bit like a dark blueish-green lance. This Italian heirloom is also known as **Lacinato** and by the very fanciful name of **Dinosaur kale**—not because it gets huge (it's the usual two-foot size) but because the pebbly surface of the leaves reminded someone of a dinosaur's skin. If you get resistance from your children, maybe you can entice them with the name. **Tuscan Baby**

Leaf is an heirloom from Italy, very quick-growing (twenty-five days) with a tender texture and mild flavor. If you've been turned off by the earthy taste of other kales, you might give this a try.

A new variety with a smaller, compact size that works especially well in containers is called **Green Curls**. A very attractive plant, with rich green leaves that are tightly curled and frilly looking, it is also bred for tender taste. **Dwarf Green Curled**, an heirloom from Scotland, is very similar in appearance. Compact size (twelve inches), easy to grow, and very hardy—an ideal choice for containers.

And by the way, if you ever see potted plants for sale in the autumn with pretty, frilly, colored leaves and labeled "Ornamental Cabbage," those are kales.

MUSTARD

That little jar of yellow mustard in your refrigerator was created from seeds produced by a plant with edible green leaves that have a little peppery tingle to their taste. Yes, mustard greens and yellow mustard are all from the same plant.

Like the other leafy greens in this section, mustard is a cool-season crop. If you start with seeds, you can plant as early as a month before your spring frost date, and then more every four weeks up until about a month before your hottest summertime days. Then, plant for a fall crop at the very end of summer, if you wish. They tolerate cold weather just fine; in fact, many people think they taste better with a little frost. Like the others, mustard greens can be harvested when still young for baby greens; they also respond well to cut-and-come-again snipping (see page 94).

An All-America winner for its tender texture, high yields, ability to tolerate both hot and cold weather, **Green Wave** is also extremely pretty: bright-green leaves with very frilly edges that look for all the world like lace trim.

It isn't bright red, and it's not gigantic in the way you might expect, but **Red Giant** is a real winner. It grows quickly (twenty days for baby greens), doesn't mind cold weather, is slow to bolt in hot weather, and is delicious as both baby greens and spicier mature leaves. It's also strikingly beautiful: the broad, flat leaves are a rich burgundy with a bright-green midrib.

There are two aliases for **Tendergreen**: first, **Komatsuna**, in honor of its Japanese origins, and second, **Spinach Mustard**, because of its value as a substitute—it doesn't bolt in hot weather. The flavor is milder than most other mustards, but like them that flavor intensifies as the plant matures. In the kitchen they can be treated like spinach when harvested young and like cabbage when more mature.

SPINACH

You don't need me to tell you about the sterling qualities of fresh spinach. You already know it's good for you, you know just how you like to prepare it, and you know how to sneak it past your children.

All the general information about leafy greens described at the start of this section applies to spinach, but there are a few extra things you need to keep in mind. My general recommendation that you use transplants wherever possible still stands, but spinach is a bit finicky about transplanting. It also germinates readily in cool weather, which is when you will be growing it. So for spinach I suggest a double-barreled approach: put in a few transplants whenever you find them, and *at the same time* sow seeds directly into your container. You'll be covered if some of the transplants fizzle out, and your harvest time will be stretched out.

The other consideration is that spinach is more sensitive to temperature than the other plants in this section. It simply will not survive hot weather, but it absolutely thrives in cool. So plant your seeds as soon as possible after your spring frost date, and then make another planting a few weeks later, along with any transplants you find. In late summer, you might want to sow more seeds for fall spinach.

Because spinach is so well established in our lives, seeds for the classic varieties are widely available. When you start researching, you'll notice two distinct groups of spinach: those with smooth leaves and those whose leaves have a crinkly, ruffled surface, described as *savoyed*. If you had a notion to plant a huge market garden at a friend's farm, you'd like to know this information, because the savoyed leaves are more difficult to wash. But unless you just like knowing things for the fun of it, you probably won't care for your container garden, because the taste is the same.

Among the longtime favorites are **Melody**, an All-America Selection; **Bloomsdale Long Standing**, an heirloom with crinkled leaves; and **Indian Summer**, which as you might guess from the name is a good choice for fall crops. In recent years, breeders have developed varieties intended to be grown as baby greens. When they are fully mature, the leaves are still on the small side, making them perfect for containers. One such is **Baby Leaf Riverside**, available from several sources; and **Little Hero** and **Catalina**, both from Renee's Garden.

LETTUCE

Lettuce might just be the absolute perfect vegetable for container gardens. It's easy to find, handsome, and grows fast; it offers lots of choices and doesn't take up an unreasonable amount of space in your container. You probably serve it several times a week, and everybody likes it. It's way better than store-bought, not least because you can gather just what you need for tonight.

Lettuce Basics

- This is very definitely a cool-season plant.
- It can be successful in either full or part sun.
- Start with transplants wherever possible.
- Looseleaf varieties are easier to manage; there are many wonderful choices.
- In cool, wet weather, keep an eye out for slugs.

PLANTING AND GROWING TIPS

Pick up transplants at the nursery just as soon as they're available, and plant them four or five inches apart in a good-sized container. Lettuce plants don't need deep soil, so if you happen to have a container that is broad but not deep, this is a perfect use for it.

Like all leafy vegetables, lettuces will thank you for steady watering; the leaves take on a bitter taste when the soil dries out. Not as essential as watering but not a bad idea: a bit of high-nitrogen fertilizer occasionally.

You may face two problems: slugs and bolting. Slugs, I'm sorry to say, like the cool and damp climate that lettuce plants thrive in. They will eat holes in the leaves—not attractive but not fatal. Check page 86 for controls.

Bolting is another problem entirely. It's the term for that sudden, astonishing vertical growth that a lettuce plant undergoes when the weather turns hot. It seems to happen overnight, and there's no fix. Once a lettuce plant has bolted, the leaves are inedible; dig the whole thing up and toss it. You can delay the inevitable by choosing varieties that have been bred for bolt resistance. There is no such thing as zero bolting, but some varieties can handle heat better; look for phrases like *slow bolting* in catalog descriptions.

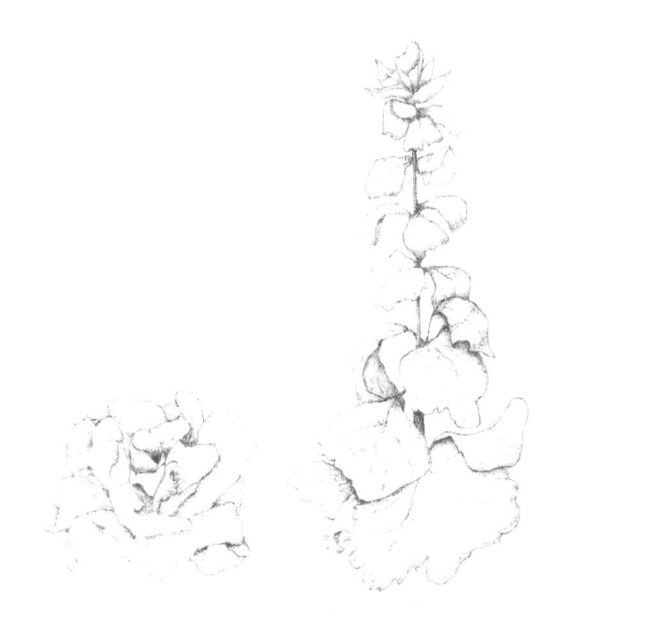

This is what bolting looks like. On the left, a healthy looseleaf lettuce, minding its own business. On the right, that same lettuce when hot weather has caused it to bolt; it seems to happen almost overnight.

HARVESTING

One of the sterling qualities of homegrown lettuce is its ease of harvesting. If you have chosen looseleaf varieties (see below), you simply pull or snip off as many of the outer leaves as you need for tonight's salad, and leave the rest for another day. Another bonus: lettuce readily lends itself to the magic of cut-and-come-again harvesting (see page 94). You simply grasp a handful of whatever amount you need for right now, slice that bundle off an inch or so from the soil, and the plant will regrow new leaves from that spot.

Varieties

If there's any downside to growing your own lettuce, it's that there are so many choices it's easy to feel overwhelmed. To make life easier, focus primarily on the looseleaf types. You already know what looseleaf lettuce looks like; you probably bought some in the supermarket not long ago. If not, the name is completely descriptive: rather than forming into a ball or tight bundle when mature (think iceberg lettuce for a visual example), the plant produces individual leaves that are loosely grouped around a central core but not otherwise connected to one another. This means you can readily harvest just a few leaves at a time, so you get more use out of your limited space.

Among the many wonderful looseleaf varieties are some longtime favorites. **Salad Bowl**, with its pretty lime-green leaves, was an All-America winner in 1952 for its tender texture and resistance to bolting, and today it's still the favorite of many home gardeners. Now there is a red-leaf variety, cleverly named **Red Salad Bowl**; the two are stunning planted together. Another All-America choice is **Red Sails**, with beautiful bronzy-red leaves; you can harvest them as baby greens as early as two weeks from seeding. In this general category, a special delight is **Lollo Rossa**, with gorgeous coloration and a compact shape with an overall soft, frilly look. The multicolored leaves look like the joyous work of some celestial gardener with a grand paintbrush; they start out green at the bottom and gradually blend upward into a dark wine-red.

And now, with miniature versions of romaine and butterhead lettuces available, it is possible to expand our horizons beyond the looseleafs. I'm sure you are familiar with romaine lettuce: long, slender leaves in a tight bundle with a vertical shape; they are best known for their crisp texture. Butterhead lettuces are quite different, with low-growing rosettes of smooth green leaves with a remarkably tender texture; you might have noticed them in the supermarket, packaged individually in round plastic tubs, sometimes with the original root ball still attached. The miniature varieties are well worth some of your garden space—delicious gourmet treats that are cute as all get-out.

I grew up in Massachusetts in the tiny community of Auburn, near the city of Worcester. Before the war, my parents owned a restaurant in Auburn called Dutchland Farms. The war shortages and rationing caused the restaurant to go out of business. Thereafter, my father worked in a defense factory of some kind in Worcester.

During the war years, the farmer next to us let us use some of his land for free, and like many people, we planted a Victory Garden. The space was about fifty by fifteen feet, I believe, and part of it was under a wonderful apple tree. We had baked apples a lot.

In the garden we grew carrots, peas (Mother's favorite), radishes, lettuce, and I'm not sure what else. My father did most of the work in the garden, but he taught me how to help, and I loved it. I was allowed to do some of the harvesting, including cleaning a few carrots just for me. But what I mostly remember was the bunnies that came to visit our garden, and my father scaring them off with rifle fire.

He would sit at an upstairs window early in the morning, where he could clearly see the bunnies, and fire off several rounds into the sky. Up to that point I never even knew he owned a rifle, and I was really upset. My dad sat me down and explained that he was very careful and never actually hit any of the rabbits, just scared them. And the nearest neighbors were a quarter mile away, so he never worried about stray bullets. Now, looking back, I realize that losing any of the vegetable crops to those little bunnies could have been serious for the family. The rabbits seemed to like the lettuce best of all, but it was probably important to keep them away from everything. Still, back then I liked bunnies much better than lettuce.

I'm eighty-six years old now, but when I think of that garden, I remember it as being a good thing to do. What great memories!

Dorothy Dodd Allen
Lynn Haven, Florida

Two mini butterheads are **Tom Thumb**, from several sources, and **Garden Babies** from Renee's Garden. They handle hot weather better than most lettuces, so you can grow them through the summer in most areas. In both cases, the whole thing is no bigger than a baseball. Imagine the amazing individual salads you could create. For miniature romaines, look for **Little Gem**. This charmer has been around for twenty years but is still cherished for its delicious gourmet taste. Because it is so popular and well established, I think you have a good chance of finding transplants in the springtime. The mature plants are about four inches in diameter at the base and six inches tall. High heat tolerance makes this another good choice for summer lettuce.

LETTUCE MIXES

So yes, there are about a million different varieties of lettuce, but fortunately there is one very easy and fun way to experiment with several: just about every seed house and most garden centers sell a mix of different lettuces, all in one packet or small nursery pot.

If you're planting mixed seeds, you never know exactly what will come up where; that's part of the fun. In the garden center the mixture will be labeled something innocuous like *lettuce blend* or *spring lettuce* or just plain *lettuce*. You'll see that it's a mass of different types of plants, but you may or may not be able to recognize individual varieties; they're all packed in pretty tightly and perhaps not showing enough of themselves for you to spot the differences. Doesn't matter. When you're ready to plant, tip that mass out of the pot, gently separate the individual plants (a brief soak in a tub of water helps), and tuck them into your container; pretty soon you'll be able to recognize what you have. Or not; maybe it turns out you have a few things you've never tried before. That, too, is part of the fun.

MESCLUN

And from the same sources (seed house or garden center), shelved along with the lettuces you can also find a product called *mesclun*. That's not the name of a particular type of lettuce plant. It's the name of a special salad blend: several different lettuces *plus* other greens, such as spinach, or red mustard, or golden chard, or mizuna; or some of all those, carefully curated with an eye toward color as well as taste. It's a wonderful way to try different flavors, even if you can't put an exact name to each individual component. I predict you're going to develop a reputation for spectacular salads.

PEAS

The arrival of spring is symbolized in many ways, depending on one's perspective. There are robins for bird lovers, daffodils for poets, cherry blossoms for Japanese families, and opening day of Chinook salmon season for Northwest anglers. For me, it's measured in a handful of sugar snap peas picked five minutes ago.

If your only experience with fresh peas is what you purchased at the supermarket, you haven't really experienced fresh peas. I'm not kidding. The natural sugars in the peas that make them so amazing begin to turn to starch the moment they are picked; by the time they hit your market, most of the true flavor has dissipated. Only people smart enough to have some in their garden can enjoy the full glories of the real thing.

CAUTION: NOT SWEET PEAS

Sweet peas, those lovely, old-fashioned garden flowers with the heavenly fragrance, are not the same as the peas in this chapter. Even though they look similar, sweet pea flowers are *not* edible. Plant them for their beauty, but don't eat them. And make sure to coach the kids too.

On the other hand, the flowers of the pea plants in this chapter *are* edible, and they make a pretty garnish—except it means sacrificing that future pea that will never happen.

Here I have grouped together three different kinds of peas, because their growing conditions are so similar. But it's important that you understand the differences among the three versions so that you end up with just what you wanted and not one of the others.

Standard green peas, sometimes called *English peas*, are the most common. When you buy a bag of frozen peas, this is what you'll have (unless the package is specifically marked as one of the other two). They are ready to harvest when the pods are full (because the peas inside are plump), but the shells are tough and inedible. You shell the pods, save the peas, and throw away the shells. If you're ever uncertain about which one you have, try eating the whole thing and you'll know immediately.

Snow peas are familiar to everyone who has ever eaten at a Chinese restaurant. The technical name is *edible-pod peas*—because you eat the whole thing, pod and all—but *snow pea* has become the generic name for the whole group. These are harvested when the tiny peas inside are still immature and the pods are flat.

Sugar Snap peas are in a class by themselves. The technical name for this type is *snap pea*, but Sugar Snap, which was the name of the first cultivar, has become the generic name for the whole group. Technically these are also edible-pod peas, because here again you eat the whole thing, but the peas inside are allowed to reach a full size so that the entire pod is plump and crisp, not flat. In this way, they have some of the characteristics of both green peas and snow peas but in an incredible blend that is way more than the sum of the parts. So revolutionary that the original Sugar Snap pea was awarded an All-America gold medal, a rare honor bestowed only once or twice in a decade.

One other thing to keep in mind: in all three groups you will find both pole and bush varieties. You can grow either type, as long as you are prepared with the necessary supports. It just means you need to pay attention when you choose your seeds.

Pea Basics

- These cool-season plants thrive in springtime.
- They grow best in full sun.
- Start with seeds planted as early as possible.
- Read the seed packet carefully. Is it a pole or bush type? Are you buying the type you wanted?
- In any case, organize your support system first.

SUPPORT YOUR PEAS, PLEASE

I'm talking about physical support, although no plant I know will turn down emotional support like a friendly greeting. The urge to climb is built into the DNA of all peas, even those that have a compact, bushlike shape. The same is true of beans, but unlike green beans, which climb by wrapping their entire vines around a thick pole of some kind, peas climb by grabbing onto a vertical support with the tiny tendrils at the top of each vine. That tendril, delicate as sewing thread, has three short "arms," each one facing a different direction. Constantly, so slowly that the human eye cannot recognize it, that three-way tendril revolves around until one of those arms smacks against something vertical, at which point the tendril whips itself around that vertical structure so tightly it holds up the entire vine. It is one of nature's most amazing tricks.

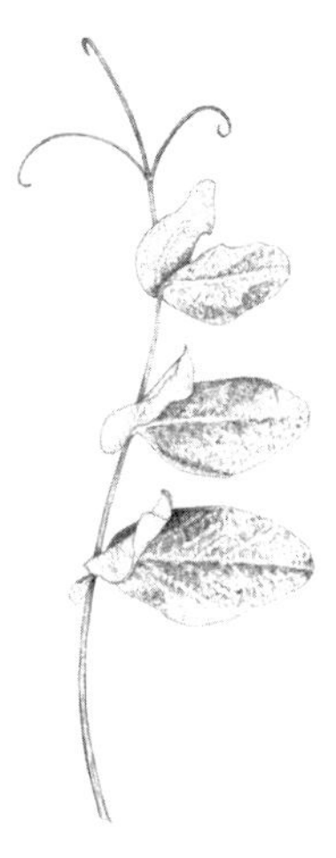

That tiny three-way tendril at the top of the pea vine, no thicker than sewing thread, will wrap itself around a vertical support with enough strength to hold up the entire vine. Amazing engineering design.

Now, for bush varieties, which I mostly suggest here, some kind of support structure is not absolutely essential. But it's not a lot of trouble, and you'll get much better results: greater productivity and easier harvesting, because the pods are easier to see. I've included several easy, inexpensive suggestions. The one thing to keep in mind is that however that structure is constructed, its vertical elements must be thin. Otherwise, the threadlike tendrils will not be able to grab onto them. And whatever system you decide on, the next steps will be much simpler if you acquire and position the support *before* planting the seeds.

Five kinds of pea supports. Remember: the vertical parts of whatever you use must be really thin. The first four suggestions are for bush peas, and should be at least two feet tall. Pole varieties need extra strength and height, at least five feet. In all cases, plant the seeds right up against the bottoms of your structures.

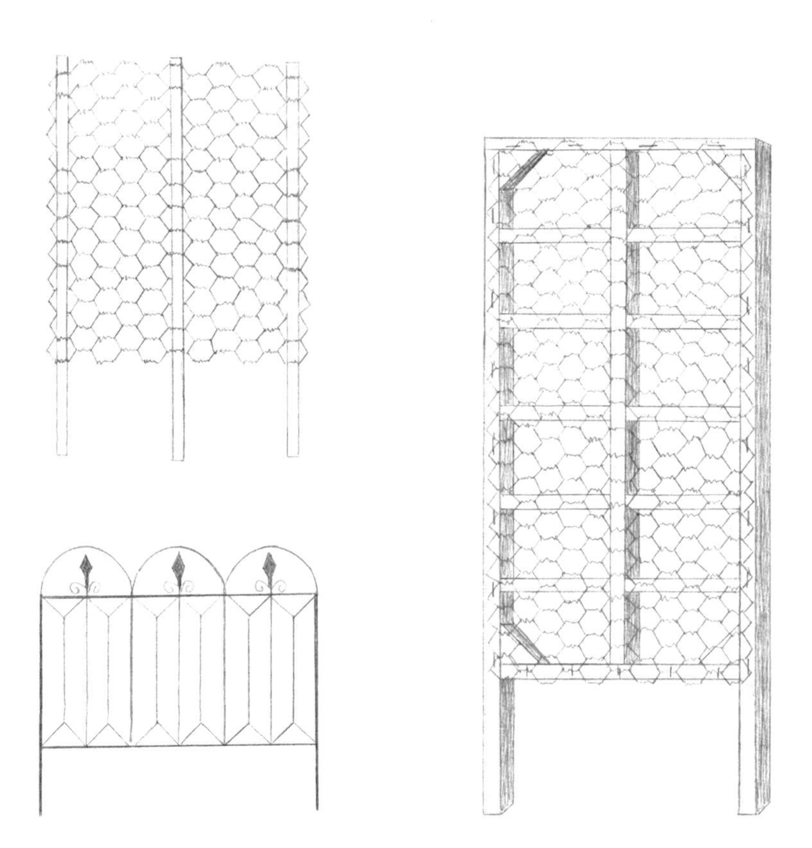

Three trellis styles for peas. *Top left*: Chicken wire threaded with wooden dowels as legs can be shaped to fit into a container. *Botton left*: Decorative wire fencing panels, designed to mark the edges of garden beds, are a good height for bush peas. *Right*: A very sturdy trellis, strong enough to support vining peas, is overlaid with chicken wire that the tendrils can easily grab.

1. From the hardware store, get a short length of chicken wire and make a little fence for one of your large containers, either crossways or around the edges. Chicken wire is not sturdy enough to stand up on its own, so while you're at the hardware store pick up a few wooden dowels and thread them through the openings in the wire, to serve as legs for your little fence. Plant seeds on both sides of the fence. Alternative: At the store, ask if they carry hog wire or something similar (larger openings, heavier-gauge wire); that material will stand up on its own, and you can make a good fence quickly. If you're not familiar with either type of wire, check the drawing in the Tomato section, later in this chapter.
2. At the garden center, look for decorative wire fencing panels intended to define the edges of garden beds. Often the panels are connected with hinges so they can be folded for easy storage, but when opened up they can be positioned in various ways. Fashion a sort of circle, or use three panels to set up a large Z-shape. If they come as individual panels, use whatever configuration best fits your planter.
3. The familiar tomato cages work fine too. Just remember that the pea plants are going to be tightly attached, so if you plan to use the cages for actual tomatoes later in the summer, you'll want to take up the entire cage at the end of pea season and clean off the remains.
4. A simple and charming solution is a small forest of branches trimmed from a tree or shrub, stuck in no particular pattern in the spot where you plan to plant your peas. You want thin branches with a small diameter and ideally with lots of smaller side branches and twigs as well; strip off the leaves. As the peas grow, they will cover up the branches so that they seem to stand without any support at all.
5. If you have your heart set on a pole type, you will need a very sturdy trellis. My suggestion: purchase a freestanding trellis at the garden center or home improvement store, something tall and rigid (or sweet-talk your favorite carpenter into building one), and cover it loosely with a network of chicken wire or fishing line.

PLANTING AND GROWING TIPS

Start with seeds direct-seeded into your container soil. Fair warning: you may find transplants in your garden centers mid-spring. Resist. In my experience, they just don't work nearly as well, and you will have committed some of your garden space to disappointing results. It's so much easier to start with seeds in the first place.

Sow the seeds approximately one inch deep and one inch apart. Do this as early as possible. Often in garden manuals you will see the phrase *as soon as ground can be worked*. That means when it is no longer frozen solid or waterlogged, usually about a month before your spring frost date. And here's a tip: if you soak the seeds in plain water overnight just before planting, they will germinate faster. Not essential, but getting a small head start is usually a good thing.

In your garden, peas are largely trouble-free. They like consistent watering, especially once they start flowering, but need little to no fertilizer. As the weather warms up, be on the alert for powdery mildew (see page 87); your best preventive is to keep the leaves dry by watering carefully (wet the soil, not the plant) and doing it early in the day.

HARVESTING

The "when" is different for each of the three main groups. Just remember, once the plant hits full growing mode, things happen quickly. The tipping point between perfectly ready and too tough can be short indeed. My best advice is to teach yourself to recognize what a perfectly ripe pea looks like by tasting a few samples when you think they're ready; you'll know right away if you were too early or too late, and then you'll know what to look for going forward.

- Garden peas (aka English peas) are ready when the pod overall is fat and firm; if you imagine it in cross section, it would be almost round. To double-check, open a pod and taste a couple of peas.
- Snow peas are ready when the little peas inside are just beginning to show the ghost of outlines on the pods; the pods themselves are flat. Check the catalog or seed packet descriptions; most will tell you the usual mature size, measured as length of the pod.
- Sugar snaps at harvest stage look very much like English peas—firmly plump pods bursting with crisp sweetness. Here again, teach yourself to recognize the "I'm ready" look by tasting a few samples.

The "how" of harvesting is the same for all, because they all have those little tendrils that grip whatever is nearby with unbelievable strength. If you simply grasp an individual pea pod and pull, you are quite likely to pull up the whole plant. Instead, use the two-hand system. With one hand, hold the main stem of the vine from which you plan to take a pea pod. Use the other hand to remove the pod, either by pulling carefully up and away or by snipping with your fingernails or clippers (which is safer).

In 1944, we had moved into a rental home in Ponca City, Oklahoma, and my parents were planting their first Victory Garden. I begged to help, so my mother gave me a bag of onion sets to plant down one row. They were surprised when I immediately came back asking for more, until they discovered that I was putting them all into the same hole.

At about the same time, our landlord, who maintained his own garden on the property, asked my parents if they knew what was happening to his strawberries. Had they noticed the birds eating the berries? Yes, a three-year-old bird named Sara. I was often found in his patch, red juices on my face and dirt on my knees.

But the real wonder was the amazing garden managed by my grandmother, Cora Mackey. She lived in Blackwell, Oklahoma, about twenty miles away. Her garden was nearly the entire backyard, full of every kind of vegetable and with flowers on the perimeter. She did all the work herself, having been widowed in her forties (she was in her sixties at this time, and lived to ninety-one). I especially remember thinking how lush all the plants were. Tomatoes, potatoes, green beans, beets, peas, lettuce, and more were overflowing rows that she had laid out with a stake and string to keep them straight.

Several times a week she would fill a wicker basket and walk to neighbors to gift them with her bounty. She also canned many things for wintertime, and her cellar was a delight of color with all the jars of fruit and vegetables lined up. I remember snapping green beans with her and shelling peas to get ready for the canning process, but I wasn't allowed near the actual pressure cooker.

Sara Chase
Vancouver, Washington

Varieties

Garden (English) peas are so firmly established in America's kitchens that seeds are widely available. As long as you pay attention to trellis needs and get those seeds in early, I think you'll do well with just about any of them. Here are two with good reputations. **Mr. Big**, an All-America winner, is aptly named; its pods easily reach at least four inches in length, often more. The AAS judges liked its high yield, disease resistance, and superior taste. The vines are not especially long, two to three feet, but need some kind of support structure because what makes this variety desirable—the abundance of large pods—also makes the vines heavy. **Tom Thumb** is another with a perfect name: its dwarf size (eight to nine inches tall) makes it ideal for container gardens, and staking is not required. An heirloom variety from England, this little cutie does especially well in cool weather, even more so than most peas. If you're itching to get going in very early spring, this would be a good choice.

Snow peas are easily distinguished by their flat, edible pods. For classic Asian stir-fries, this is the one you want. Two longtime standards, still considered the favorite by most gardeners, are **Oregon Giant** and **Oregon Sugar Pod II**. Yes, I live in Oregon and yes, both of these were developed at Oregon State University by Dr. James Baggett, the legendary plant breeder, so you might assume there's a bit of favoritism in these recommendations. Nope—I recommend them because they are the best. Both have strong resistance to disease and are not as bothered by mildew as others, which makes them a good choice for late-summer crops, and both provide exceptional flavor on highly productive vines. Oregon Sugar Pod II vines are about thirty inches long and need a trellis for top results; mature pods are about four inches long. Oregon Giant lives up to its name; the mature pods are about five inches long and one inch wide. In return for a good trellis, the three-foot vines will give you a vigorous crop.

Snap peas are a true marvel. Even little kids who claim to hate vegetables love to snack on them. Because of their popularity, many admirable varieties have been developed to meet this demand, and you should have no trouble finding good seeds from multiple sources. Just a tiny reminder: pole varieties will need a tall, sturdy trellis. Bush varieties can be grown without support but do much better with it.

Let's begin with **Super Sugar Snap**, an improved version of the original Sugar Snap that knocked everyone's socks off. It has strong resistance to the common pea diseases, including mildew; tolerates heat better; and offers greater productivity and more manageable vines (five feet long). Note, though, that five-feet vines definitely need a trellis.

Among the many bush varieties, you can't go wrong with any of these: **Sugar Daddy**, with vines twenty-four to thirty inches long and loaded with three-inch stringless pods, has strong resistance to disease and mildew. **Little Crunch** is similar in size and has the same excellent productivity and disease resistance. **Sugar Sprint**, also of similar size, produces a vigorous crop extra-early and also tolerates heat well, so it's a good choice for both early-spring and early-fall crops. **Sugar Bon** is even smaller, with vines reaching twelve to twenty-four inches and prolific yields of three-inch pods; it's excellent for both spring and fall plantings. A strong favorite for many years is **Sugar Ann**, beloved for its compact size (eighteen to twenty inches), for the fact that its mature pods hold on to the vines for up to a week without turning tough, and for its supersweet taste. All-America Selections, describing this 1984 winner, particularly noted its early productiveness, "up to two weeks earlier than comparisons."

Let's close this rich selection with a brand-new idea: a dwarf variety that produces plump, edible pods that look like a green bean but taste like a snap pea. It's called **Snak Hero**, an All-America Selection in 2020, and it features a perfect size for containers (eighteen-inch vines) and a succulent, sweet taste. Best of all for folks looking to make the most of their limited space, it's a stunner when grown in hanging baskets.

RADISHES

By now, you know one of my favorite things about growing your own vegetables: the joy of having, right outside your door, special treats you will never find in the supermarket. That's even true of radishes, as you will see shortly. But there's an even simpler reason to grow them in your container garden: they are so doggone satisfying. In as little as three weeks, you can go from seed in the soil to radish in the salad, without a lick of effort on your part (except maybe watering). Success is a powerful thing.

Radish Basics

- Start with seeds. Always.
- Don't forget about thinning.
- Full sun is best; partial shade is OK.
- They thrive in cool weather.
- Be aware of the different types: spring, summer, and winter.

THE SEASONS OF RADISH

Like many other vegetables that prefer cool weather, radishes can be planted in early spring up to the onset of warm temperatures, then again in late summer for a fall crop. But the world of radishes gives us an extra bonus: varieties specifically for winter, in addition to the familiar spring-then-fall planting regimen. And in between the spring and winter types is one more: summer radishes. Almost a full year of wonderful options. To get the results you want, be sure to pay attention when buying seeds and put the various types in the soil at the right times.

PLANTING, GROWING, AND HARVESTING TIPS

In taste and texture, the various types of radishes share many similarities. But there's enough difference in the gardening techniques to warrant individual descriptions here.

Spring radishes. Starting as early as possible (when the ground is no longer frozen), plant seeds about one-half-inch deep. Once the leafy tops are established (which could be as little as one week), thin the seedlings to a two-inch distance; this is important. They grow amazingly fast, so you may want to sow more seeds at two-week intervals. And perhaps another batch or two at the end of summer, for fall harvests.

As a group, these radishes are super easy to grow: essentially disease free, all they really need is consistently moist soil. So if it's not your rainy season, you'll have to step in with regular watering. This is the singular key to well-shaped, fully formed radishes with perfect texture.

Harvesting is essentially a matter of paying attention: most spring types are at their peak, firm and crisp, when about an inch in diameter, and many will reach that size in as little as three weeks, so keep checking. All the seeds you planted in the same batch will mature at approximately the same rate, so just check one and you'll know about them all. Choose one, brush away the soil at the base of the foliage, and gauge size by eye. If it seems the right size, pull it up and squeeze. If you get a firm resistance, that's good; but if your random sample feels soft, it has gone past the ideal stage and is heading toward unappetizing—pithy, hollow, tough, and unpleasantly hot in taste. If left in the soil at this point, this fellow and its neighbors will only continue to deteriorate, or they will bolt in the hot weather. Or both.

Summer radishes. The key quality that separates summer radishes from their springtime cousins is that they have been found to tolerate warm weather better. In most other ways, the gardening techniques are similar. Plant the seeds in late spring, thin the small seedlings, and be vigilant about keeping your soil moist. Consistent watering will be even more important in the warm months, and you may also have to be ready to provide some shade on long, hot summer afternoons. (You'll know—the green tops will start to wilt.) They take a bit longer to mature than spring radishes, and the "ready" size is a bit larger than spring types but not nearly as large as the winter varieties, just comfortably in between.

Winter radishes. The notion of wintertime radishes may be new to you, but I think you will find them intriguing and well worth exploring. Note: This is not the same as planting more of the familiar spring radishes in the late summer for a fall crop. I'm talking here about something completely different and pretty darned terrific.

Winter radishes are, in many respects, like the other types you've known for years: colorful, crisp embellishments to your salads or snack trays. The difference—and it's a big one—is in the way they grow. Whereas spring and even summer radishes grow quickly and seem to deteriorate just as quickly if left in the garden too long, the winter types are just the opposite. They grow more slowly (sixty to ninety days rather than twenty to thirty), reach a much larger size, and can stay planted until you need them (or until the soil freezes solid) without bolting or turning woody or pithy; they just keep on getting larger and, some think, sweeter. And even when dug up just ahead of a hard freeze, they retain good quality in the refrigerator or a cool storage area for weeks.

Sow seeds a half-inch deep in late summer, six to eight weeks ahead of your first fall frost date. Thin the seedlings to two-inch spacing, and harvest some at baby size. Thin the others to four-inch spacing, and they will keep on growing in that space as long as your soil doesn't freeze completely. Harvest one or more whenever the mood strikes, and then start paying attention to the weather forecast. When a hard freeze is on the way, quickly harvest everything; trim off the leaves, wash and dry the roots, wrap them in plastic, and they will wait happily in the refrigerator (or root cellar, if you're lucky enough to have one of those).

Varieties

SPRING TYPES

You can find seeds for those little red darlings in garden centers, hardware stores, practically everywhere you turn around, and, of course, at many online sources. There are numerous variations, pretty much all the same, but why not start with the classic: **Cherry Belle**, named an All-America winner in *1949* and still everyone's favorite more than seventy years later.

When you're ready to expand your palette a bit (the color palette), check out **Plum Purple** from several sources (where it is sometimes labeled **Purple Plum**), a really luscious rich magenta color; and **Pink Beauty**, a bright, hot pink that at the moment seems to be even more widely available. The color is in the outer skin; all these are white on the inside.

Perhaps the best way to enjoy the beautiful colors—certainly the easiest—is to take advantage of the collections put together by several seed houses, in which seeds of different varieties in a range of colors are combined into a curated mixture; each packet contains some of all the seeds. One nice mixture, widely available, is named **Easter Egg** (white, red, pink, purple radishes). Baker Creek has a similar mixture they call **Easter Basket**, which contains some long, skinny types as well as the familiar rounds; and Park Seed offers **Beauty Blend** (white, yellow, magenta, and very dark red). You can't tell what color any radish is until you pull it up—and that is the fun of it, especially if you have small children around.

SUMMER TYPES

One summer type you may already be familiar with is **French Breakfast**, a French heirloom from the nineteenth century sometimes marketed today as a gourmet treat. The flavor and texture are the same as familiar spring radishes. The growing conditions are much the same also, except that they take a bit longer to reach harvest size (thirty to thirty-five days). The difference is their shape and unusual coloring. They are oblong, about the size of your thumb, bright red most of the way down to a pure-white tip. Today, many people enjoy them as cocktail hors d'oeuvres, with a dab of fresh butter at the tip then dipped in salt.

Another summer type, not quite so well-known, is **Icicle**. They are well named: snow-white in color and long and slender in shape. And though they are often assumed to be a subvariety of Daikon radishes, which are also white and oblong, these are different: more pointed, slenderer, sweeter, milder. Similar, but with a shorter length, is **Mini Mak**, from Johnny's. And if summer weather finds you craving the red round radishes from springtime, you might like to try **Giant of Sicily**, an heirloom type from Italy. They are just like their smaller cousins in every way except larger, about the size of a Ping-Pong ball.

WINTER TYPES

Now, for the sheer fun of it, I'd like you to meet **Watermelon** radishes. These round beauties can be enjoyed within thirty days at their baby stage or stay in the ground up to hard freeze, when they'll reach their mature size, three to four inches in diameter. The outer skin is an unremarkable tan color, but the inner flesh is a brilliant rose-red, just the color of a perfectly ripe watermelon, with a reliably crisp texture and tender flavor that is sweeter and less spicy than most other radishes. They're a beautiful surprise for winter salads and appetizer trays, and a fresh delight all the way around.

Another nice winter type is **China Rose**, sort of the reverse of Watermelon in coloration: luminous pink skin and white interior. Long and cylindrical in shape, these Chinese heirlooms offer a crisp texture with a peppery undertaste. At mature size, they are sort of like fat, chunky carrots: wider at the shoulder (as much as two inches in diameter), tapering to a more slender tip at a full length of six inches or so.

A HANDY TRICK WITH RADISHES

Radishes win the gold medal for growing quickly, which makes them perfect for introducing children to the joy of gardening and pretty darned satisfying for everyone else. But you can also turn that trait into a very useful garden trick with several applications.

- Mark the space where underground plants are planted but not yet showing.
- Delineate a space you are reserving for peat-pellet transplants with a "living fence" of radish plants.
- Intermix with slower-germinating seeds, such as carrots, to keep the soil loose and mark the spots.
- Fill in around plants that are just starting to show themselves and don't need all their allotted space. All cool-weather vegetables are perfect for this—peas, maybe. But you can also tuck a few summer radishes in beside warm-weather transplants, like tomatoes. By the time the radishes are ready to harvest, those other veggies will be ready for the extra space.
- Remember the container gardener's motto: no bare dirt. Wherever there is even a tiny empty spot in a container, tuck in a few radish seeds. It will make you feel virtuous.

TOMATOES

A popular country song reminds us, as only country music can, of the deep-down value of the "only two things that money can't buy"—true love and homegrown tomatoes. Now, your love life is your business and I wouldn't tip even a single toe over that line, but I'm pretty sure we can all agree on the second thing. There is simply nothing like tomatoes fresh from the garden, dead-ripe and dripping with the essence of summer, especially in comparison to the cardboard replicas we so often find in the grocery stores. So I'm not going to spend any time convincing you to include them in your garden; I'll just give you a few pointers on tomato success, and leave you and your true love alone to enjoy the results.

Tomato Basics

- Start with transplants wherever possible.
- These are hot-weather plants that crave full sun; at least six hours a day is your goal.
- Tomatoes need lots of water and a strong root system; provide room for both by planting deep.
- With just a few exceptions, you need a deep container and at least twelve inches of soil.
- Learn the key distinction between the major categories.

ONE ESSENTIAL THING

See that last point in the list above? It is absolutely critical—and absolutely foreign to new gardeners. So let's get it out of the way first.

All tomatoes fit into one of two very broad groupings: *determinate* and *indeterminate* (a new variation, really a subgroup, is called either *semi-determinate* or *semi-indeterminate*, depending on one's perspective). The difference lies in the way they grow, and it has a big impact on container gardeners especially. You can grow either type as long as you know in advance which you have and have arranged for the necessary supports.

Determinate. These tomato plants grow to a certain height range (built into their genes) and no more. (I think of them as being "predetermined.") A familiar example: Roma types, with the approximate size and shape of an egg, popular for salsa and cooking. They set flowers on the tip ends of their branches, pretty much all at the same time, which means the fruits ripen pretty much all at the same time. This makes them especially appealing for container gardeners because there is much less need, sometimes none at all, for staking or trellises. The trade-off is, there are fewer varieties available.

Indeterminate. These tomato plants grow in a steady, continuous pattern: a little bit of growth on the main stem, then a few side branches, which lead to a few blossoms, which develop into a few fruits. Meanwhile a new short growth section develops on the main stem, with new side stems and new blossoms—then another little bit, and another, and so on and on, until cold weather puts a halt to everything. This means the plant will give you ripe tomatoes a few at a time, slowly, over an extended season, rather than all in one concentrated period. Botanically these are vines, and they will grow to the roof of your house if the weather stays warm; if it doesn't, they are just as likely to leave you with a bumper crop of green tomatoes.

This lusty way of growing means that indeterminate plants need some type of physical support. The trade-off here is you will have many more options to choose from. Many of the longtime favorite cherry tomatoes fall into this category, like the much-loved Sweet 100, which you may already know.

Semi-indeterminate. Now some very good news for container gardeners. In recent years the tomato universe has expanded to include a new grouping called semi-indeterminate or semi-determinate, depending on which end of the spectrum you start from. In this happy middle ground, plants show the indeterminate growth habit (gradual production over an extended period) on vines that are more like determinate plants, much shorter than the usual indeterminates. The net result is a smaller, more compact plant that is much easier to manage in containers (although most will still benefit from a modest support of some kind) *and also* a steady supply of sweet treats. The best of both, really.

A few years ago I would not have been able to introduce these new varieties to you.

By the time you read this, there will probably be many more. It's just one more example of how significant container growing has become in the world of horticulture. What was once considered silly is now widely accepted and wildly popular. So if anyone ever gives you a hard time about your decision to grow a container garden, hand them a few tomatoes and smile sweetly.

THREE WAYS TO SUPPORT YOUR TOMATOES

Indeterminate tomatoes, those without a built-in growth definer, will need you to provide some kind of support structure: a wire cage, tall stake, or trellis. If you don't, they will sprawl all over the place, leaving the poor tomato vines to fall on the floor, defenseless against marauding squirrels and your cousin Harold's big feet. And you need to have those supports ready at planting time. Here are three easy possibilities; retail and online garden supply companies offer more options.

Three tomato supports, left to right. A sturdy tree branch, with trimmed stubs to keep the fabric ties from slipping. A standard cage from the hardware store. A strong cage made from heavy-duty wire.

1. **Tree branch.** If you have access to a strong branch from a tree or large shrub, convert it into a rustic tomato stake. Remove all the side branches, but leave short stubs for each; these stubs will prevent your ties from sliding all the way down. Grown this

way, with one main stem and multiple side branches, the plant is easy to maintain and ripe tomatoes easy to reach.

2. Tomato cage. You may have noticed these at the garden center or your local hardware store; they are a very common sight in late spring. They work well for plants on the small side but can be a bit too flimsy for really vigorous growers. Gardeners sometimes manage by doubling up, one inside the other.
3. Heavy wire cage. Ask your friendly hardware store manager for a length of wire with a heavier gauge and bigger openings, and make your own cage. You might start by asking if they carry hogwire, popular with farmers and fence builders, or remesh, used by concrete contractors. If not, show them this sketch; you want something sturdy enough to stand up on its own and with openings large enough you can get your hand through.

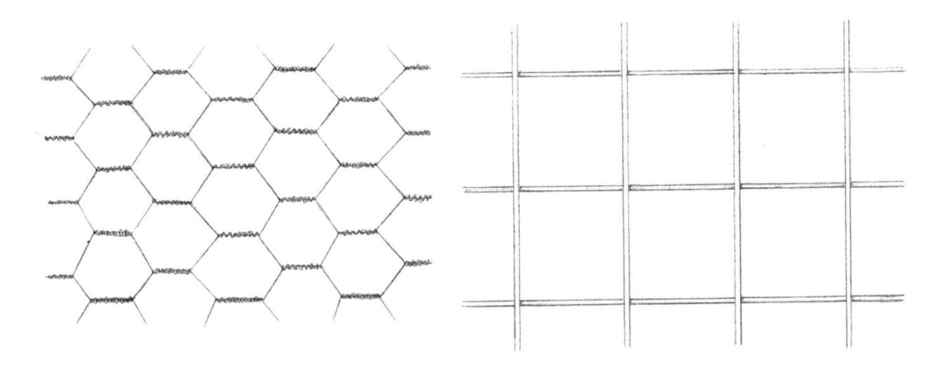

Chicken wire (left) is familiar and widely available but relatively flimsy. Hogwire (right) is sturdy enough to stand up on its own. Both would make good custom-size support cages for things like tomatoes, but chicken wire will need to be supported with strong stakes.

PLANTING TIPS

Just about every gardening book, magazine article, blog, and tip sheet you will ever encounter will say the same thing: when adding a new plant into the garden (container or otherwise), plant it at the same soil level it was in the nursery pot. And that is true.

Except for tomatoes.

Tomatoes need three things for success: sun, lots of water, and a strong root system. Two of those are in your direct control, and you accomplish both with the same technique:

deep planting. The process is really quite simple, even though it does feel a little weird the first time you do it.

First, tip the tomato plant out of the pot it came in from the nursery; if it doesn't come easily, give the whole thing a brief soak in water. Count the number of side branches on your plant, from the very bottom to the very top. Divide that number in half. Starting from the bottom, completely remove that many side branches; snip them off all the way back to the main stem. Now you have a very weird-looking tomato plant, with a few healthy side branches at the top half of a long, otherwise bare, stem.

Next, in the biggest, deepest container you have, dig a deep hole; it doesn't have to be wide, just deep. Gently position your plant at the bottom, spread the root ball out a bit, and then evaluate: the lowest set of side branches should be about one inch above the soil line. If you're not sure, lay a short stick across the hole and note where it touches the tomato stem. Adjust, if needed, by filling in the hole or digging it deeper. Then fill in around the plant with the dirt you removed to make the hole, plus some of your extra potting soil if needed, tamp everything down carefully, and water the whole thing well so the plant snuggles down in its new home. Allowing for the settling, the bottom branches should now be about two inches above the soil.

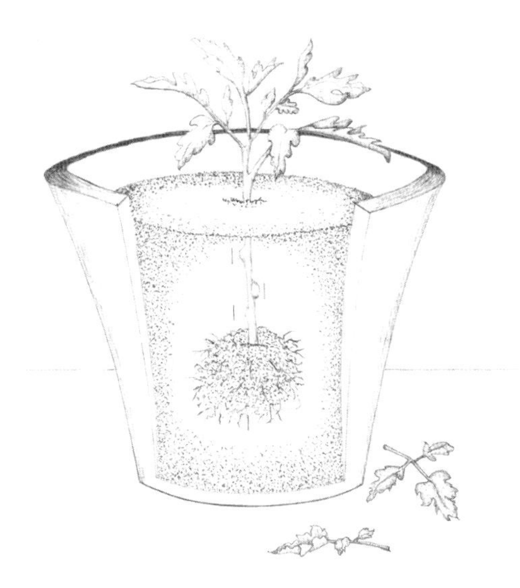

To help your tomato plant develop the full root system it craves, snip off half of the existing branches and plant the whole thing deep in the pot. New roots grow from the nodes where branches used to be.

Why did you do this to your beautiful plant? To encourage a vigorous root system. Because here's the magic: all along the buried stem, where the side branches used to be, the plant will grow new roots from the nodes (that slight bulge where all the growing energy is concentrated). And that, in combination with the original root ball, will give you the robust root system that is the key to everything.

If what you've just planted is an indeterminate tomato, you have one more task. You should already have at hand whatever support system (stake, cage, or trellis) you have chosen. Now put it in place, right up against the tomato. It's much easier to get it in position now, while the plant is still tiny and forgiving.

GROWING AND HARVESTING

Water. Tomato plants need a lot of water (which they take up through their roots, which is why a good root system is so important). And they do most of their growth in the summertime, when your weather cycle is probably at its driest. So, my friend, it's up to you.

You remember how to check for soil moisture, don't you—stick your finger down as far as you can. If it's dry all the way down, add water. Water the soil deeply, taking care not to get the plant leaves wet (that can promote a fungal disease), then don't water again until things are really dry. Your goal is a consistent level of soil moisture, not irregular bursts of too much water after a too-long dry spell; that can cause the tomatoes to split.

Fertilizer. A dose of fertilizer every couple of weeks, particularly when the plant starts setting fruits, is a good idea. Use either a balanced, all-purpose type or one formulated for vegetables (low in nitrogen). Since you'll also be vigilant about watering, this would be a perfect spot for my plain water / fertilizer water two-step dance (see page 85).

Possible problems. Tomato plants do occasionally come with a few problems, but we are willing to forgive because the benefits are so glorious. Still, here are a few things you should know about. One is blossom end rot, a term so descriptive you'll know it when you see it: a big, black mushy circle where the flower originally was (directly opposite where the tomato is now attached to its stem). The most common cause is uneven watering, which we already know you won't be doing. Another problem is cracking or splitting of the outer skin; it's unsightly but not fatal in itself, except that it opens the door to other diseases and unwelcome bugs. And guess what—it's also caused by uneven watering, so you know how to avoid it.

Speaking of bugs, keep an eye out for these two. Cutworms, which hide out under the surface of the soil, will slice the plant right off at the soil line, like felling a tree. You'll never see them at work, just the poor dead plant lying on its side. Remove the corpse, put on some gloves, and poke around the spot for big brownish worms like caterpillars; if you find any, grab them and head straight for the trash. The other possible visitor is a tomato hornworm, with its fat, striped body and distinctive horn. They, too, are hard to find, because they are

I was born in December 1941, right after Pearl Harbor, and my parents had just built a home next to my father's parents in Alvadore, Oregon. My grandparents had moved to Oregon in 1925, after trading their wheat farm at Khedive, Saskatchewan, for the property. Both households always had a dedicated garden spot, war or no war, for vegetables fresh in the summer and canned to last during the winter.

Our garden was in an old barnyard that was paneled off to keep the cows out as they went from the barn to the pasture. We had a couple of trellises of cane berries on one side, then rows of carrots, onions, lettuce, cabbage, spinach, cucumbers, both winter and summer squash, sweet corn, and my father's tomatoes, for which he had a special technique. Dad had a contract with the army to supply 1,600 dozen eggs each week, and for that he maintained 3,000 laying hens. Of course that also meant lots of chicken manure, which he put to good use. For each tomato, he would dig a large hole, three feet by three feet, and fill it with chicken manure before adding the tomato plant.

Once, when I was quite young, I went into what we called the *straw shed* and found several deer hanging to be butchered. There was lots of game in the area and several local farmers had organized a hunting party through a wooded area on our farm and harvested the meat to feed their families. This meant they could sell the beef, pork, and chickens they were raising to the companies providing supplies for the war effort. Those fellows are all dead now, so I guess we can't arrest them for poaching.

Bill Sanborn
Portland, Oregon

tomato-leaf green, but at the first sign of big holes chomped in the leaves and the tomatoes, put your gloves on and go hunting.

Harvesting. This is mostly a matter of waiting patiently until those beauties are truly ripe, and that is mostly gauged by color. The only time you might have trouble is at the very end of the season, when you're trying your best to let the green tomatoes stay on the vine as long as possible, to catch the last of the warm days before the nights get so cool that the plant just shuts down. (The critical point is forty degrees; when nighttime temperatures are above that, you can risk it one more day.) If the ones you're hoping to save are still fully green when cold weather hits, they won't ripen; start researching recipes for fried green tomatoes. But if they are showing even a blush of color change, it's worth trying to salvage them. Take them inside, into a warm, protected spot, and cross your fingers. Some people have good results tucking them inside a paper bag.

But whatever you do, don't keep ripe tomatoes in the refrigerator. The skins get tough, the insides get mushy, and most of the flavor is lost. This applies to all tomatoes, including the ones you buy from the store, but especially for those jewels you worked so hard to grow. Keep them on the counter, perhaps in a pretty white bowl to accentuate their colors. And don't be surprised if the little ones tend to disappear, a handful at a time.

Varieties

Tomatoes are, hands down, the number-one favorite for home vegetable gardeners. That undisputed fact means, among other things, that there are more choices for tomatoes than for any other vegetable, and more new ones show up every time you blink. At this moment, there are more than a thousand varieties; by the time you get ready for your garden, that number could be much higher. How can anyone possibly decide which to grow?

Let me see if I can help you sort things out. Here are a few general rules of thumb.

1. Concentrate on determinate or semi-determinate types, especially if this is your first container garden. Veteran gardeners know there are good reasons for choosing indeterminates, but they tend to take extra attention. If you're interested in a specific variety for which both determinate and indeterminate versions are available, stick with determinates; you'll be much happier. Next year, you can branch out. Note that

My name is Susan Novak, of Carrollton, Texas. I inherited my love of gardening from my parents, Anthony and Dorothy Novak, both of whom learned from their own parents. I visited my mother at her home in Austin, Texas, and asked her to tell me about my grandparents' Victory Gardens. Her story is shared below.

I was born in 1925 and grew up in Chicago, near Wrigley Field. Across from our alley was a vacant lot, and the man who owned it divided out ten-foot garden plots and invited neighborhood families to plant anything they liked. My father, William Reinking, was raised on a farm, and he chose mostly tomatoes for our garden. My brother and I were given the job of picking off the green hornworms; we both hated it. We were also given the responsibility of watering the tomatoes and watching over them until they were ready to be picked, and that was much nicer, especially because they were so delicious.

During the war years, Americans were encouraged to plant a Victory Garden to grow their own food, so our family garden became even more important. Anything extra was shared with neighbors, and men would come by weekly in a truck and sing out to sell fruits and vegetables to those who needed more than they could grow.

But it wasn't just my family. Everyone had a Victory Garden, you know, including your father's family in Michigan. Joseph and Helen Novak, his parents, my future in-laws, had immigrated to the US in 1907 from Poland and settled in St. Clair. They always had a garden, partly because they loved it and partly because it produced enough food to sustain their large family of seven children through the year. From that garden came potatoes, kohlrabi, cucumbers, tomatoes, and many other veggies. Helen canned everything she could, for the winters, and their large basement was filled with bins of root vegetables and rows of canning jars. When we visited your Novak grandparents, you and your sisters were often sent down to the basement to collect something for dinner. To young girls, the basement was a dark, scary place, filled with ghosts and monsters. But you survived, and eventually enjoyed playing in the basement.

Dorothy Reinking Novak
Austin, Texas
As told to her daughter

in the lists that follow, I suggest indeterminates only if there is also at least one close alternative that is either determinate or semi-determinate.

2. Next, think about how you wish to use your tomatoes. That will lead you to the question of size, which is how the suggestions are organized. For snacking and fresh treats like veggie kabobs, the many versions of cherry types, with their happy colors, are wonderful. "Cocktail" tomatoes are just the right size for salad slices and wedges. For cooking or sauces, choose one of the Roma types, and medium or large slicers will make your sandwiches shout hallelujah.
3. Then, look for special features that have meaning for you. For instance, some varieties are particularly suited to hanging baskets, which would be great if your garden space is really tight. Some, described as *early*, are better for short growing seasons. Some, often called *dwarf*, have particularly compact foliage growth, thus also good for limited spaces. And just for the joy of it, how about colors other than red.

As always, I'm suggesting specific varieties to get you started with your wish list, but I encourage you to include alternates, too, just in case.

CHERRY TOMATOES

Here we encounter one more example of just how goofy the popular terminology in the plant world can get. On one hand, "cherry" is often used to refer to *all* tomatoes of the small, one-bite size. But within that large category are several subcategories, each given the name of a familiar fruit it resembles in some way. Thus we have grape tomatoes, with the shape (but not color) of really large seedless grapes; currant tomatoes, teeny little things about the size of actual currants; and pear tomatoes, the general shape of pears, with a small top and fat belly, but in a smaller size. And guess what? There is also a cherry subcategory called *cherry*, with small tomatoes the shape and approximately the size of actual cherries.

Why does this matter? Because if you explore seed catalogs as a way to familiarize yourself with additional varieties, which you know by now that I encourage you to do, you will find tomatoes described with this cherry/grape/currant/pear terminology, and those who write the catalog descriptions assume you already know what those terms mean. Now you do.

"Cherry" cherry tomatoes will give you the widest selection. Many are indeterminate, not my first choice for you unless they have other strengths, but fortunately we have good possibilities for more compact varieties that are easily managed in containers. Let's start with **Patio Choice**, in two versions, **Red** and **Yellow**. Both are determinate plants specifically bred for containers, and **Yellow** was named an All-America Selection winner in 2017. Maturing in fifty to sixty days, both produce scads of sweet little tomatoes, about one inch in diameter, on short vines. If you have the space, planting one of each together makes a striking container.

Next, several charmers that are classed as dwarf determinate or semi-determinate (operationally, the same thing). **Tidy Treats** will give you a full season's supply of sweet red treats, rich with what some have described as heirloom flavor, on a plant that, true to its name, will always stay compact and tidy. (Those nice folks at Johnny's Seeds call it a "polite" growth habit. Polite tomatoes—I want some.) Another fine semi-determinate is **Lizzano**, chosen an AAS winner in 2011 for its admirable taste, high yields, and growth habit—the vines naturally cascade downward, which makes this one especially good for hanging containers. Delicate, bright-red tomatoes begin to appear after about sixty days and flourish all summer long. The vines are short, only twelve to sixteen inches, and maintain an overall compact shape that the catalog writers at Territorial Seed call "well mannered." Another brilliant descriptor; I can't do better. And then there is **Litt'l Bites**, from Renee's Garden Seeds. Another semi-determinate with a cascading growth habit that seems to call out for a beautiful hanging basket, this one is known for its high productivity of small red fruits with a supersweet taste. Among its many fans is Deborah Miuccio, who runs the test gardens at Gardener's Supply (see appendix). "This is the most compact variety of tomato I've ever grown," she says. "It doesn't get bigger than two feet tall. I harvested over five hundred cherry tomatoes from a single plant."

Let's close out this cherry-cherry grouping with one indeterminate tomato that I couldn't bring myself to leave out. **Sungold** is, for many, the gold standard (no pun, really) of cherry tomatoes, with an unparalleled sweetness and a gorgeous color somewhere between polished gold and fresh-made orange juice. The plants are usually between four and six feet tall, and are very vigorous growers, meaning that they need very strong support. Remember that indeterminates are large and gangly, and along with a sturdy cage or trellis will absolutely need your attention. Many, many gardeners over many, many years have concluded the effort is worth it.

Here are two nice, well-behaved **grape cherry tomato** varieties for your consideration. **Ruby Crush**, sixty days, will make any container sparkle with richly colored red tomatoes dangling in clusters. Best of all, these plants are determinate and need little support, often none at all. Many gardeners are excited about **Celano**, seventy days, a new semi-determinate so exceptional it was an AAS winner in 2020. One judge praised its sweet flavor, pleasing texture, deep color, and "phenomenal yield." It grows larger and taller than Ruby Crush, and so will benefit from a tomato cage or something similar.

Note for **currant and pear-style cherry tomatoes**: as far as I can tell right now, these two types are available only as indeterminates. So because there are no alternates to suggest, they are not included. Let's hope some "semi" varieties appear soon.

COCKTAIL SIZE

We might think of these as cherry tomatoes' big brothers. The shape, coloration, and sweet-tart flavor are very similar, but cocktails are bigger, about the size of a Ping-Pong ball. They are very meaty with few seeds, which makes them perfect for adding to salads. Here are a few for your consideration, all of them determinate.

Let's start with one that offers a promise in its very name: **42 Days** says it all. This one is remarkably early, and it also is not intimidated by cool weather. So you can plant sooner and enjoy the sweet red tomatoes way ahead of your friends with traditional, in-ground gardens. Serious bragging rights await. **Red Racer,** fifty-seven days, was named an All-America Selection in 2018, with particular praise for its flavor and high yield. Plants can reach as much as three feet in height, so you might want a tomato cage. Finally, **Patio,** seventy days, so compact it is sometimes referred to as a dwarf determinate. The branches (vines) grow closely together, laden with tightly spaced dark green leaves that form a beautiful backdrop for bright red tomatoes. Take note: this is not the same variety as Patio Choice, described above.

ROMA TOMATOES

Roma tomatoes have thick, meaty flesh, few seeds, and very little juice. These are the ones you want for cooking with tomatoes, such as for your grandma's famous spaghetti sauce, or if you enjoy canning (their low water content is a plus here). All Romas are the approximate shape and size of eggs, usually red, and almost all of them are determinate, so you can do well with just about any healthy transplant you snag at your favorite garden center. But here are a couple suggestions, all of them determinate, for your handy wish list.

An heirloom variety from Italy, **Martino**, seventy days, is especially noteworthy for its amazing productivity: literally hundreds of tomatoes from one plant. In fact, keep a close watch as soon as the first tomatoes start turning red, because ripe ones have been known to fall off to make room for the new fruits right behind them. Plants can sometimes get as big as three feet, so you might want a short stake. **Paisano**, sixty-eight days, is described as "a San Marzano style," meaning it is very similar to the famed Italian tomatoes in both flavor (wonderfully sweet) and shape (a bit longer and thinner than most Romas, with a pointed tip). Want to impress your foodie friends? Serve them pasta with Bolognese sauce you made from your garden and tell them they're eating San Marzanos. And wouldn't it be fun to have a golden Roma? **Sunrise Sauce**, fifty-seven days, high yielding and low-maintenance, will reward you with a rich crop of unusually sweet tomatoes in a lustrous color that glows golden-orange in the sunshine and tastes like heaven.

MEDIUM TO LARGE SIZE

Until just a few years ago, big slicing tomatoes, the kind you want for your world-famous, straight-from-the-grill hamburgers, were almost certain to be indeterminate. Fortunately for beginning container gardeners, there are now other options.

The strong, compact **Bush Early Girl** plant is fully determinate, about two feet in height, and will give you loads of large, four-inch slicers that are bright red and extremely flavorful in fifty-four days, which definitely qualifies as early for tomatoes. Take note: this variety is a compact form of the classic Early Girl, hugely popular with backyard gardeners ever since it was introduced in the 1980s. Because of that continued popularity, you will undoubtedly see Early Girl plants in your garden center starting in late spring; but do *not* buy one unless the tag clearly says it is the bush version; the other one, without the word *bush* in the name, is indeterminate and, trust me, you probably don't have room for it.

Another determinate tomato that produces slicers in a short time (well, short for tomatoes), **Early Doll**, fifty-five days, delivers shiny red rounds with a stellar flavor. The plant itself can grow up to as much as four feet, so it will thank you for some support, like a standard cage.

The semi-indeterminate variety **Super Bush**, seventy-two days, from Renee's Garden Seeds, produces large red slicers, about four inches in diameter, with a wonderful sweet flavor. The mature vines average a manageable three feet, but because they are heavy with large tomatoes, it's a good idea to use a cage or short stakes.

An All-America Selection winner in 1984 and steadfastly popular ever since, **Celebrity**, seventy days, is technically categorized as determinate but is often considered semi-determinate because it just keeps on pumping out its round red fruits all summer long. The plants reach a height of three to four feet, with tomatoes of about a four-inch diameter, a wonderful all-purpose size. In fact, many gardeners think of Celebrity as "the little black dress" of tomatoes—meaning it works for almost everything. And a classic choice for almost forty years? Hard to argue with success.

THE LAST WORD

A vegetable garden in the beginning looks so promising and then after all little by little it grows nothing but vegetables, nothing, nothing but vegetables.

—Gertrude Stein

CHAPTER 6

The Good Stuff: Herbs

If you do nothing else, treat yourself to a small container garden of herbs. That one container, overflowing with fresh herbs right outside your kitchen door, will immeasurably brighten your life. How wonderful is this little herb garden? Let me count the ways.

- Most herb plants tolerate going dry and may even prefer it, so they won't punish you if you forget to water them. And most would rather you *not* do any fertilizing.
- They are almost always well-behaved; a small, compact growth pattern is built into their DNA. This is extra-good news for container gardeners.
- Harmful insects don't like them; in fact, many gardeners deliberately plant herbs as a repellent.
- Most of them are perennials, which means they come back every year with no extra effort on your part.
- If you've never cooked with fresh herbs, you're in for a grand surprise.
- They don't require a lot of maintenance, but when you do work on them, your hands smell wonderful afterward.

There are many ways to define herbs, but most people think of them in the context of cooking, and that will be our focus here. And, for all sorts of practical reasons, I'm reluctantly limiting this chapter to the culinary herbs I believe most of you will be interested in and for which seeds/plants and information are readily available.

MEDITERRANEAN HERBS IN YOUR GARDEN

It turns out that many of our most popular culinary herbs are native to the same part of the world: the area right around the Mediterranean Sea. Here the weather is hot almost year-round and very sunny, with low rainfall, and the soil is mostly sand and rocks—not exactly fertile. So native plants tend to have small, tight foliage (the opposite of lush) in which the aromatic and flavorful properties are condensed.

This is all good news for container gardeners. It means that these herbs will tolerate low watering just fine, and vigorous fertilizing is actually counterproductive. To say this another way, I don't know of any other group of plants that are so happy if we just leave them alone.

BASIL

Lovely basil, favorite of many cooks, is also a gardener's delight. Because it is such a handsome thing, with so many wonderful varieties to explore, we forgive it for being picky about weather and needing our attention for watering.

Basil Basics

- Unless you live in the tropics, this is an annual (meaning it lasts only one season).
- Transplants are recommended where possible, but starting from seeds is easier than you might think.
- Basil insists on hot weather.
- It prefers full sun.
- Regularly "pinching" stems will curtail legginess and encourage more leaves.

GROWING BASIL FROM SEED

In general, I think first-time container gardeners are better off starting with basil transplants. But if circumstances call for starting with seeds, as they sometimes do, you will find that process much more manageable with basil than with most other plants.

The process is remarkably simple because you're going to direct-sow right into the big container—no transplanting needed. When the weather has warmed up to the point that nighttime lows are consistently above fifty degrees, plant a few seeds about a quarter-inch deep and one inch apart into their final container home. The seeds will germinate in seven to ten days. When they have two sets of true leaves, thin your little forest down to the strongest one (or several, if you're aiming for a mixed-basil container). The others go into the compost bucket or, if you have the time, pot them up into small containers and gift them to gardening friends. The young plant grows so quickly that you will be harvesting leaves in as little as three weeks.

SUCCESS WITH BASIL

Your biggest challenge might be patience. Don't do *anything*—plant seeds or transplants from the nursery—until nighttime temperatures are reliably at least fifty degrees. At the other end of the season, keep a close watch on weather forecasts. Basil simply cannot withstand cool temperatures, so even the slightest hint of frost is your cue to jump into action. If the prediction is for a brief, light chill, you might keep things going by moving the container temporarily into a sheltered area or covering the plant with some kind of protective layer. But if actual frost is in the forecast (thirty-two degrees or below), basil season is over. Get out there and harvest everything now, because tomorrow morning the plants will be toast—shriveled, black, and sad.

Otherwise, the best way to enjoy a luscious crop of basil leaves is to keep the plant producing side branches with more and more leaves. You do this with a mild kind of pruning often called *pinching*, because the stems are soft enough you can do it with your fingernails. It's really straightforward (study the drawing on page 151) and extremely effective, because it is based on a simple but vital bit of botany: if you remove the growing tip from any stem or branch, the next buds lower down will be stimulated to open up and develop into side branches. And if you then pinch out *those* stems once they grow a bit and develop their own buds, two more lower buds will open, and so on and on. And of course put the trimmings to good use in tonight's pasta sauce.

Do this faithfully, and your basil plant will maintain a full, bushy shape, which is far more attractive than something gone leggy, and will give you lots more leaves, which is the whole point.

For ongoing general care, you don't have much to do. You will have to keep an eye on watering because you're growing these plants in the hottest part of the year. Fertilize lightly with a nitrogen-rich fertilizer (nitrogen promotes leaf growth, remember) starting when you first add transplants and only occasionally thereafter during the summer. Many herb gardeners believe that heavy fertilization produces large leaves in which the aromatic properties are thinned down. Your goal is to help the plants do what they naturally want to do when the sun is shining bright—and no more.

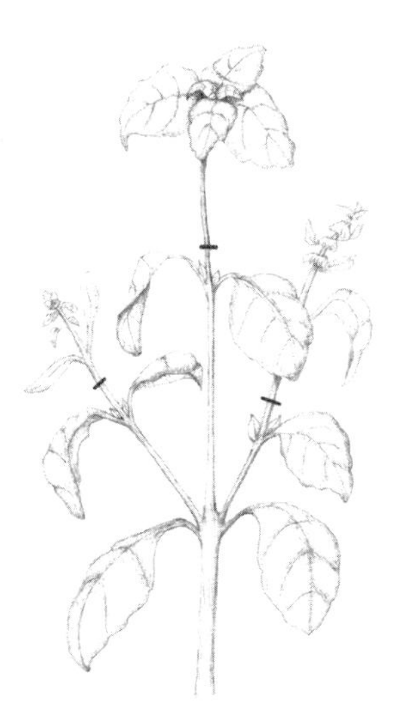

To keep a plant's shape compact and bushy, pinch off (or cut) leggy stems just above a growth bud in the V where leaves attach. Those buds will then grow into new branches, which may need their own pinching at some point.

At some point in late summer, your basil may start to form a spray of flowers at the outer tips of the most mature stems. Many people believe that once the plant starts to flower, the overall flavor of the leaves deteriorates, and so they watch carefully and religiously clip off all flower buds.

But as with all herbs, the basil flowers are edible, and many gardeners enjoy them for their own sake. They are small and dainty, and carry a whisper of basil taste along with their tender beauty. They make lovely garnishes to summertime beverages or a tossed salad, and they can turn plain white vinegar into something magical. Or you can mediate between the two camps and have it both ways: leave some flowers (bees love them) and remove others.

Varieties

If your life with basil up to now has been limited to dried green bits in little glass bottles, you are in for a real treat. From just about any good garden center, you can find a rich selection of basils with large leaves, small leaves, deep-purple leaves, and many delightful types that blend two flavors together so that you get a true basil taste with an extra layer of something else: lemon, lime, cinnamon, clove, or anise, for example.

So many, in fact, that it is easy to feel overwhelmed. It might help if we think in terms of broad categories, described below. Here, as elsewhere in this book, I offer suggestions for specific varieties I think you might like, but if you don't find that exact thing, look for something else in the same category, and all will be well.

ITALIAN BASILS

This is where you will find the plants that come to mind when you hear the word *basil*—the most common, most recognizable, most widely available. And number one in this group is the plant known as **Sweet Basil**. The leaves are a bright medium-green, slightly cupped (like a small canoe), and offer a tantalizing mix of flavors: anise, mint, and pepper blending into a complex taste we identify simply as *basil*.

But right away we come face-to-face with one of those annoying bits of nomenclature confusion. Sweet Basil is a specific plant, known to botanists as *Ocimum basilicum*. On many seed racks you will find a range of packets from different companies, all labeled simply "Sweet Basil." (And for those of you who just enjoy knowing stuff, since the genus name *Ocimum* is derived from a Greek word that means "basil," the full scientific name of this plant is actually "Basil basil.") But there are quite a few varieties derived from this "mother" plant, and all those varieties are collectively referred to as (lowercase) *sweet basils*. If you see yourself making pesto, any of these Italian varieties are what you want.

Genovese basil is probably the best-known variety of Sweet Basil. Its leaves are dark green, shiny, and pointed at the end; they are also a bit longer and flatter than Sweet Basil, with what many people consider a richer, deeper, spicier flavor. True Genovese basil originates from Genoa, Italy, but its popularity has prompted development of many new versions with similar characteristics; they are usually described as "a Genovese type." Among the possibilities is **Dolce Fresca**, an All-America Selection (2015) prized for its superior flavor and attractive, compact shape (good for containers). Another outstanding choice is **Italian Cameo**, a cross of Italian Genovese and a traditional dwarf basil from Renee's Garden. Also noteworthy are space-saving versions of two well-established favorites: **Genovese Compact** and **Prospera Compact**, from Johnny's Seeds. If you're looking for either of these two, be sure to pick the "compact" version; that word should be in the variety name.

PURPLE BASILS

Purple basils are easy to spot at the garden center: all of them feature intensely colored leaves in dramatic shades of purple and deep wine-red. That rich color is the main attraction for many gardeners, but the leaves are also very flavorful, with a distinctive licorice flavor. Popular varieties include **Dark Opal**, an All-America winner in 1962; **Purple Ruffles**, another AAS winner (1987); **Red Rubin**; and **Purple Opal**. Those with more sophisticated palates than mine could tell you how their flavors compare, but I do believe you would be pleased with any of them.

THAI BASILS

Thai basils are a distinct variety with very recognizable botanical characteristics, a perceptible anise taste, and a maddening mashup of common names. You may find, for example, plants bearing name tags like **Thai Lemon Basil**, **Thai Cinnamon Basil**, **Thai Anise Basil**, **Thai Licorice Basil**—all of them completely different from western varieties with those same names minus the "Thai."

I think you're much better off learning to spot the Thai type by its appearance, which is especially obvious side by side with standard Italian types. Thai basil leaves are smaller and narrower with a pointy tip, the flowers are pinkish-purple, and the stems are purple (not green). One way out of this tangle is to focus on well-known, long-established varieties, such as **Siam Queen**. An All-America Selection in 1997, it has been a favorite of home gardeners ever since. Another good possibility is **Queenette**, from Renee's Garden, using seeds imported from Thailand.

The characteristic flavor of Thai basils is well-known to folks who love Asian cuisine. With its strong note of anise accented with a layer of peppery spiciness, it is different from the more familiar, milder, and, some say, sweeter taste of Italian basils. Adventurous cooks might appreciate the fact that Thai types hold their flavor through high heat better than the Italian ones. But I'm pretty sure all but the fiercest purists would happily substitute one for the other if that were all they had on hand.

GLOBE BASILS

Globe basils are characterized by their shape—small, round globes about the size of a cantaloupe, with petite, intensely flavored leaves. Their dainty size is a real boon to container gardeners, as they can be tucked into even the smallest spaces and do quite well. They are also very appealing when planted all alone in a small pot (say six inches in diameter), especially if that pot is a rich contrasting color.

I believe you'd be delighted with any of those little darlings you are lucky enough to find in the garden center, but keep an eye out for (or order seeds for) **Spicy Globe**, an heirloom variety with delicious little leaves on a tight dome shape that stays compact at six to nine inches; and **Spicy Bush**, very similar except for a slightly larger size. **Windowbox Mini Basil** from Renee's Garden, an Italian import, grows into a neat umbrella shape about eight inches wide, with a thick mass of fragrant, tiny leaves.

SPECIALTY BASILS

Here, in this catchall category, I want to introduce you to a few basils that offer gardeners something special. **Persian**, an All-America Selection in 2015, has a complex, rich taste that is both spicy and lemony and all-around wonderful, but I particularly want to highlight its flowers. Late in the growing season, this basil will offer you (and all the bees in your zip code) a flurry of robust flower stalks laden with tiny blossoms that are partly white and partly a soft pinkish-purple, and so dainty they look like a big spray of orchids bred for fairies.

Double-flavored basils (my term, not in the least scientific) combine the familiar basil taste of warm spiciness with something else, something recognizable as a separate flavor layer—for example, cinnamon, clove, lime, and lemon. Lemon cultivars are becoming more and more common, and thus I believe your chances of snagging a handsome transplant at your garden center are good indeed. And if you happen to find one of these next two, even better.

An heirloom beloved by professional chefs and home gardeners for generations, **Mrs. Burns' Lemon Basil** delivers intense fragrance and flavor in a large, robust plant. The plants may reach twelve inches or more in both height and width, so it will be happiest in one of your larger containers. A nice plus if you live in the South: this variety tolerates extreme heat better than most. Relatively new and instantly popular is **Sweet Dani Lemon**, an All-America Selection (1998) described there as "greatly improved." The lemony fragrance is darn near irresistible to humans and bees but not to mosquitoes and flies—many consider it a good repellent. Like Mrs. Burns, Sweet Dani is larger than many container basils, as much as two feet tall with one-foot flower stalks, so it, too, needs generous space to do its best. Because of its tall, upright growth pattern, you might even like it as the centerpiece in a large, multi-plant container.

CHIVES

Chives are a member of the onion family and clearly earn their spot in the family tree through their taste (very like scallions) and their way of growing (from an underground bulb). Does that make them a vegetable? Maybe, but I think of them as herbs, with their tender flavor, bright green foliage, and charming aromatic flowers—and maybe you do too. Even if all you do is snip them up to garnish your scrambled eggs or steamed mussels, you'll be glad you planted some. But they offer so much more.

For one thing, they are so easy on the gardener. You plant a few seeds or baby plants, and every year they just keep going, spreading into an ever-larger patch that still remains compact and tight, politely not crowding out anything else, and giving you an almost endless supply of delicious green shoots ready for clipping. And they reward your no-care care with lovely pink flowers, tiny pom-poms that taste exactly like the essence of chives on a floral background.

Chives Basics

- These are perennials, meaning they come back every year.
- They prefer full sun.
- Start with young plants whenever possible.

SUCCESS WITH CHIVES

Chives grow from a small bulb underground; each year, those bulbs multiply, eventually producing a tidy clump of long, thin green shoots, and every year that clump just keeps getting bigger and fuller. Year after year, they will do just fine with very little attention from you.

Planting. Planting is quite straightforward. If you prefer seeds, they are widely available and not fussy. But transplants are much simpler, more efficient, and easy to find. In garden centers (and sometimes supermarket produce sections), look for small pots of healthy, young chive plants, a tight mass of green sprouts that looks like a section cut from lawn grass that is way overdue for mowing.

You can either transfer that entire clump into your container garden as is or divide it into segments. My recommendation: using a sharp or serrated kitchen knife, cut that clump into quarters, and plant the four small clumps wherever you have an empty spot, or about six inches apart from one another if you prefer an all-chive area. Why? Because at some point you're going to have to divide a large clump anyway, and it's much easier at this stage.

Ongoing care. This is super easy. Give them a drink when you're doing your normal watering, don't get too enthusiastic with fertilizer, and otherwise just enjoy their friendly presence and faithful contribution to your meal planning. Depending on the weather patterns where you live, they may die down in winter. Don't panic; the underground portions are still healthy, storing up growth energy for the spring, when you'll see a sudden flush of new growth.

Flowers, then seeds. If your only experience with fresh chives up to now has been the small bundles you can sometimes find at the supermarket, you may not even realize that these plants put on a flower show starting in late spring. But those blossoms are a major delight of growing chives in your own garden. The most common type of chives (called *common* or *onion chives*) produce charming little pom-poms in pink or lavender that sit atop flower stalks taller than the foliage, just waiting for you to come say hello.

You know who else likes to come visit the flowers? Bees and butterflies looking for nectar and serving as pollinators in the process. So the flowers, if left on the plant to dry, will hold pollinated seeds that will drop to the ground and start new plants next spring. That's called *self-sowing* (or as we sometimes say, "the plant seeds itself down"), and it is either a nice bonus or a nuisance, depending on your outlook. Either philosophy is fine with me; I just want you to be aware.

The flowers are lovely in themselves, and a fanciful addition to a small-scale flower arrangement. But they also carry the tender onion flavor of the plant, and so make a wonderful garnish for just about anything. You can either use the flower head intact or separate out individual florets (about the size of a small grain of rice). I have found them even more versatile than the whole flower. They become little nuggets of oniony taste in pretty colors, and thus work well sprinkled onto any dish that would benefit from onion flavoring and a bit of contrasting color: cream soups, poached fish, mashed potatoes, rice pilaf, baked chicken, like that.

Varieties

Onion chives (slightly nicer than the inelegant other name, "common chives") are what you are most likely to find in the garden center. And although there are several named varieties, I truly believe that for most of us, whichever one you find on the day you go looking will serve you well. The flowers will be either pink or pinkish-purple, and always wonderful.

Garlic chives are different in taste, foliage shape, and flower patterns, but exactly like onion chives in how they grow and what they need from us. I like growing both, because I like to cook and because neither one is the least bit demanding about space. The main way you can tell them apart is their foliage: flat in garlic chives, round in cross section with onion chives. Also, the garlic-chive flower head, once it appears, is quite different. Rather than one round, fluffy pom-pom, it's a cluster of tiny star-shaped blossoms, white, and not particularly photogenic. But those flowers are hearty producers of seeds, so if you leave the flower heads on the plant, next year you'll have a few million tiny new plants popping up right around this year's plant. OK, I exaggerate.

The major reason for adding this type to your garden, though, is the flavor. It's an intriguing mix of onion but also garlic. So put your good cook's imagination to work, and think of all the dishes that would benefit from those two tastes together and would also be brightened by emerald-green tidbits.

MARJORAM AND OREGANO

I'm going to do something unorthodox here and combine these two closely related herbs into one section. Why? Because most people consider the flavor to be quite similar (in fact, lots of folks can't tell the difference), and so choosing just one or the other is a reasonable strategy for container gardeners with very limited space. There are some differences, of course, which I will describe so that you can decide which one (or ones) goes at the top of your wish list.

Marjoram and Oregano Basics

- Start with transplants.
- Both need full sun and lots of it.
- They're vulnerable to root rot, so go easy on watering.
- Both are perennials but are susceptible to cold, especially marjoram.

GETTING TO KNOW MARJORAM AND OREGANO

Both are technically perennials, but both struggle in really cold weather, especially marjoram. If you want to keep them growing outdoors, mulch heavily and be prepared to add some protective cover or move the pots to a protected area in bitter cold. Marjoram in particular is classed as a *tender perennial*, meaning it is perennial only in the warmest zones. As a consequence, most North American gardeners treat it as an annual, starting over each spring with new plants. Some people have good luck with cutting their plant back rather severely in late fall and transplanting the trimmed result into a new, smaller container for a sunny indoor windowsill; after spending the winter inside, the plant gets repotted for outdoor gardening.

Otherwise, remember the basics for these Mediterranean classics: full sun and plenty of it, very light watering, very little fertilizing. Both plants put on a show of dainty, pink flowers which make lovely, savory garnishes. But mostly we grow these two for their richly flavored leaves, and so this is where the difference between the two plants might matter to you.

Different tastes? Trying to describe the flavor of two plants so closely related leads us into a thicket of "-er" comparatives. Some say marjoram is sweeter and oregano is spicier. Or marjoram is more floral, oregano earthier. Oregano is sharper; marjoram is warmer. Marjoram is subtler, oregano more robust. And so on. But everyone seems to agree that while the tastes are very similar, marjoram is milder and oregano is stronger. In practical terms, then, it's perfectly fine to substitute one for the other as long as you adjust quantities to account for the different levels of pungency.

Cooking. Cooking with these two is largely a matter of what you have in your garden. If you have both, you might find it useful to think of things this way: marjoram is more often associated with dishes whose culinary roots are in France or Britain, and oregano is more at home with Italian, Greek, or Mexican cuisines. One other difference to keep in mind: many experienced cooks prefer to add fresh marjoram sprigs at the end of cooking, to get the full benefit of its milder flavor.

Harvesting. Harvesting is simply a matter of snipping off individual leaves or stems, depending on which quantity you need. That snipping is good for the plant; it keeps it bushy and compact, with a minimum of long, straggly stems.

Varieties

First, you have a choice to make, especially if you have room for only one plant: Do you want marjoram or oregano? They are so very similar in flavor that many people consider them interchangeable. So it's reasonable to wonder, *How much does it really matter?*

My first suggestion comes down on the side of practicality: if you have room for only one and you don't already have a definite preference, forget the names and make your choice on the basis of the plants themselves. This calls for some serious hands-on research. Spend some time at the best-stocked nursery you know and study several plants of both types side by side. Especially consider their fragrance. Using different fingers, rub a leaf from each plant; the marjoram aroma is lighter and more lemony. This will help you experience the differences and decide how important that might be in your life. But always, *always* keep in mind that the differences are minor and relative, and that even professional cooks slide deftly between the two.

Your larger goal with this research trip is to find the most beautiful, strongest, healthiest individual plant, no matter which group it fits into. Or maybe two plants, if you want to try one of each.

In sum, then, my first suggestion is: find the most fabulous transplant, regardless of name. The second suggestion is to include aesthetics into your planning. Which brings us to a different type, with beautiful golden color and a low-growing, creeping habit that drapes beautifully over the edges of your container like a frilly petticoat. It is either **Golden Marjoram** or **Golden Oregano**, or **Curly Golden Oregano** or **Marjoram**, depending on which source you encounter. Once again, my take is that the name doesn't really matter. Just love it for its stunning good looks and call it Matilda if you wish.

Golden and **Curly Golden** varieties are much like each other except for the leaf design. The plain Golden types have the smooth, flat leaves of standard green versions, except for that dazzling color. In the Curly version, the edges of the leaves (botanists call them the *margins*) are wavy so that the overall look of the plant is a mass of tight curls. Along with the well-loved flavor, both add a bright splash of color to the herb garden, which otherwise tends toward perpetually green.

THE BEST OREGANO AND MARJORAM

The best oregano is the one widely known as (and labeled as) Greek oregano. And in spite of the name, this is the one you want for all your Italian dishes.

The best marjoram is the basic plant (*Origanum majorana*), probably tagged as simply "marjoram" or perhaps "sweet marjoram."

MINT

There's an old story about mint that contains, as the best stories do, a lesson worth remembering. It seems that on one of those grand estates familiar to fans of British television dramas, the lady of the manor was strolling through her exquisitely manicured gardens with the head gardener when she stopped to ask a question.

"Johnson," she said, "I've become very fond of adding mint to my tea in the afternoons. Would it grow well here? Is it difficult to plant?"

"You know how we plant mint, m'lady? We dig a hole, throw it in, and jump back out of the way."

And so do we. Even in a container herb garden, mint—in a refined, dignified way—will take over everything else. The solution is simple and brilliant: dedicate one container to mint only, and enjoy its incomparable flavor and fragrance to your heart's content.

Mint Basics

- These are perennial plants.
- Start with transplants.
- Full sun is best; partial sun/shade is possible.
- Mint spreads like crazy, so grow it in its own pot.

SUCCESS WITH MINT

Growing. The key thing to know about mint is how it grows. Underground stems grow out sideways in all directions, and from those stems, roots grow downward and new stems sprout upward. In one year, what started as a small transplant will become a nice tight mass of minty goodness. In most climates, that plant will die down in winter and arise anew come spring, quickly forming into a larger clump. Next year, larger still. And so on.

After several years, container gardeners will notice that most of the new growth is tight up against the edges of the pot. That's because the underground stems are continually reaching outward until they bump up against a solid barrier; in the meantime, the central core, representing the little transplant that you started with, is mostly bare. At that point, your best move is to dig up the entire thing, cut away the old roots, and treat the fresh new growth as if it were new transplants from the nursery. Plant them back in the center, and you're set for several more years, and thus on forever.

Maintenance. This is mostly a matter of regular watering and keeping the stems trimmed so they don't get lanky—which you'll want to do anyway as you snip segments for summertime salads, wintertime stews, and all-the-time hot and cold beverages.

Varieties

The "pure" mint taste, something you've known since childhood, comes from either spearmint or peppermint. Peppermint is not peppery, as most of us would understand that word, but it does have a stronger flavor than spearmint. If you find a charming little nursery plant labeled "mint," it's likely one of those two, and for most uses, you can consider them interchangeable.

However, that is not the full story with mints. Because mint flavor is so widely used in the world's cuisines, several dozen specialty types have been developed over the years. Most of them are a blend of two tastes: mint, which is dominant, and something else in the background. That something else then lends its name to the name of the new variety—Banana Mint, for example. Here are some of the many delightful possibilities:

- apple (also known as woolly mint)
- banana
- basil
- bergamot
- chocolate
- ginger
- grapefruit
- lavender
- lime
- margarita
- orange
- pineapple
- strawberry

Any one of those would give you great pleasure in return for little effort, and a chance to add distinctive flavors to your meals that you cannot get from just-plain-mint alone.

APPLE MINT

PARSLEY

You stop at your neighborhood sandwich shop for a quick lunch and treat yourself to their famous gussied-up BLT and legendary house-made potato chips. When it arrives, off to the side is a lone sprig of green curly something, trying in vain to hold its own against the decadent chips. Do you eat it? Nope. You might push it around on the plate, but nothing more.

Poor parsley, vivacious green and full of vitamins, but almost never chosen. Maybe there's a better way to think about this. Let's try stepping sideways. That poor neglected sprig at the lunch counter is curly parsley, also known as *common parsley*, which because of its leaf texture is not especially pleasant raw and not all that easy to cook with (difficult to chop). Serious cooks often describe the flavor as mild or bland, depending on their mood that day. But there is another kind of parsley, one with a richer flavor, all the good-for-you vitamins, and a leaf shape that is much easier to manage in cooking. I'm talking, of course, about flat-leaf parsley, also known as Italian, and it has pretty much taken over the parsley world for cooks.

To be clear: I am not advocating that we consider curly parsley a has-been. It is still the king of green garnishes, for the traditionalists among you, and a quite beautiful plant that decorates garden borders and other empty spots with lush greenery. But we're working with a few guidelines here, and one is that for container gardens, with their very limited space, plants have to earn their spots. Also, for this chapter we are focusing on herbs for their culinary value, and in that respect flat-leaf is clearly the winner.

Parsley Basics

- Start with transplants.
- Grow as an annual.
- Parsley does well in cool weather.
- It thrives in either full sun or part shade.

PLANTING, GROWING, HARVESTING, AND COOKING

Planting. Parsley is not intimidated by cool weather, so you'll begin seeing transplants for sale early in spring. Plant them in your standard potting soil, in a container that can handle a mature height of approximately one foot, and in a spot that gets at least partial sun. Keep it nicely moist, and get ready to start harvesting as soon as there are several healthy stems.

Growing. Are you getting the sense that parsley is easygoing and undemanding? You'd be right. It needs more water than the more familiar Mediterranean herbs, but as long as you don't allow it to become waterlogged, parsley is one of the most trouble-free plants in your garden. It will keep going right through spring, summer (except in the very hottest climates), and fall, when it dies back to the ground.

Or seems to. Next spring, you are quite likely to see new growth appearing from last year's corpse, and you might reasonably conclude that your parsley is a perennial. Reasonable, but wrong. In botanical terms, parsley is a *biennial*, meaning it takes two full years to go through its entire life cycle. Year one, lots of good green leaves; year two, a ridiculously tall main stalk with a few side stems, a flower head, and eventually seeds. Unfortunately, the second-year leaves are so bitter that most people consider them inedible, and so parsley is usually treated as an annual. Savvy gardeners dig up and discard the plant at the end of the first growing season and start over next spring with a new baby.

Harvesting. It's simple and logical: snip off a few stems, as many as you need for what you're cooking, and strip off the leaves. Good news: parsley is a cut-and-come-again plant (see page 94 for details). To maintain a pleasant shape, harvest the most mature stems (from the outer edges of the clump); new stems will begin to grow to replace them.

Cooking. To use parsley in your cooking and enjoy its full benefits, you might find it helpful to think of it on two simultaneous levels: a lively green garnish that also adds its own flavor in layers of bold, citrusy, herbal spiciness—one of those hard-to-describe but easy-to-love complexities we so often find in the world of herbs.

Use that lively garnish generously when cooking; it adds a happy surprise to roasted vegetables, rice pilaf, fresh green salads, soups (minestrone especially), and egg dishes like frittata. If you have more than you can use at the end of the season, make a batch of green goddess dressing for holiday gifts. And if you've never tried making tabbouleh, now is the time. It's amazing, and it absolutely depends on fresh parsley.

Varieties

The distinction between curly and flat-leaf parsley is immediately apparent when you see them side by side in the nursery, even if both are labeled simply *parsley*. And since I strongly recommend you start with transplants, the nursery is where you'll be making your choices.

For curly parsley, pick up the healthiest, handsomest transplants you can find. Tight curls of foliage on compact stems, forming an overall frilly ball of deep green—that's what you're looking for. For flat-leaf, you may come across several with variations of the word *Italy* in the name. One widely available is **Giant of Italy**, where *giant* is more about the leaves than overall height; these beauties usually top out at twelve to eighteen inches or so, perfectly manageable in your containers.

In 1942, my mother planted our first Victory Garden near Vancouver, Washington. At harvest time, she sent me through the woods with a basket of cucumbers for our neighbors. When they opened the door, I proudly announced, "I brought you some pickles!"

I think the neighbor must have relayed that to my mother, because she reprimanded me for calling them pickles instead of cucumbers. How was I to know the difference? I was only four!

Mary Funk
Chehalis, Washington

ROSEMARY

Beautiful rosemary might just be the epitome of Mediterranean herbs. It loves full sun and hot temperatures, doesn't mind poor soil, and absolutely relishes seaside climates. Its very name gives us a lovely clue: the English word *rosemary* is a transliteration of the Latin name *Rosmarinus,* meaning "rose of the sea." Planted anywhere near the ocean, within a marine climate, it grows like gangbusters. Even inland, this wonderful plant is so easy on the gardener that it will give you years of pleasure in return for very modest effort.

Rosemary Basics

- Start from transplants.
- This is a perennial.
- Grow in full sun.
- It's sensitive to very cold temperatures.

GROWING AND ENJOYING YOUR ROSEMARY

A healthy rosemary plant needs two things: evenly moist soil (neither dried out nor waterlogged) and protection from deep cold. For the first, make sure you use quality potting soil with good drainage to prevent root rot; overwatering is the number-one reason rosemary fails to survive. At the same time, don't let the soil dry out completely. For the second, use a container that you can easily move to a more protected spot whenever the temperature drops to twenty-five degrees.

Ongoing care. Aside from those two concerns, ongoing care is ridiculously easy. You don't really need to fertilize, and very few of the familiar pests will show up; the strong oils in the needles are actually a grand deterrent. Prune occasionally to keep a tidy shape, if you wish, and if you live where deer like to snack on your garden, spread the trimmings around your vulnerable plants.

Harvesting. Harvesting is easy and fun. Mature branches will be somewhat woody at their bases, so if you want a full stem, use your clippers. The new growth at the outer tips is softer and more pliable; you might be able to snap off a short stem by hand. In either case, you should do this on the same day you plan to use them, because the fragrance and flavor of fresh rosemary are incomparable.

Cooking. Cooking with fresh rosemary is a gift to yourself and your family, especially if everyone is accustomed to the taste of the commercial dried version. But it's really important that you get familiar with the physical nature of the plant's foliage. I say *foliage* because rosemary does not produce anything resembling leaves, as most of us understand that term. The foliage takes the form of short needles, very much like a fir or spruce tree, about an inch long and tight up against the stems. Except for the very newest growth, the needles are stiff and tough, becoming more so as the individual branches mature during the growing season. Most people find it unpleasant to eat a needle in a cooked dish unless it is chopped into very fine bits. The woody stems are even worse—and potentially dangerous, with the real possibility of someone choking.

So when cooking with this herb, plan on using one of these two methods:

1. An entire sprig (one or more) with needles intact, which will be removed before serving.
2. Very, very finely diced needles blended into recipes that contain some kind of liquid, to soften the rosemary bits.

And one more word of advice: Rosemary is strong and easy to overdo. Go slowly.

But with those cautions ever in mind, you will find many uses for this wonderful herb. Entire sprigs (method 1) add their magic to many soups and stews and to the braising liquid of your favorite meat dishes; slow-cooking in liquid extracts the flavor, and afterward the used-up stem is easy to remove, just like you do with bay leaves. Then expand on that basic idea. I often put a large sprig of rosemary in the simmering water when I poach chicken breasts. Or how about a crisscross of long sprigs underneath oil-tossed potatoes and carrots in the roasting pan.

To add rosemary to a dish with the expectation that the herb will be eaten, you will need to cut the individual needles into very fine dice (method 2). If your knife skills are not up to that, mix the rosemary needles with some other dry ingredient the recipe calls for (salt, maybe), and buzz them together in the food processor.

In late summer and fall your rosemary plant will very likely put on a flower show, with loads of small pastel-blue blossoms tight against the stems. Like all herb flowers, they are edible and carry the essence of their plant's flavor in a gentler, softer taste that adds a special note when used as a garnish. I especially like them sprinkled onto fruit salads or lemon sorbet.

Varieties

Speaking broadly, rosemary plants belong to one of two basic categories, defined by their shape: creeping or upright. Most serious cooks believe the upright ones have more robust flavor, but those plants can easily reach five feet or more.

Creeping types are usually a better fit for container gardeners. You don't have to be concerned about them getting too tall; they only grow out sideways, and with a little encouragement from you, they will send their new growth downward over the edges of the pot, leaving a big swath of the container free for other plants. Among creepers, **Prostratus** is the classic; another good possibility is **Huntington Carpet**, with tighter, denser foliage.

Among the upright types are several that don't get overwhelmingly tall: **Boule**, which grows into a three-foot dome shape; **Collingwood Ingram**, which grows wider than tall (four feet by two feet); and **Ken Taylor**, with a growth pattern like Collingwood but trailing and a bit shorter. For petite size all around, there's **Blue Boy Rosemary**, with tiny needles and even tinier blue flowers; it may eventually reach a mature size of two feet tall and wide but will take several years to get there. These are worth a look if you live where weather patterns are unfriendly to rosemary—that is, cool or wet or both: **Salem**, two feet tall and wide, and handles wet soil well; **Arp**, well-known for tolerance of cooler temperatures; and **Hill Hardy**, three or more feet tall and wide, and can survive zero degrees. Two others that make a cook's eye sparkle are **Tuscan Blue**, long cherished by professional chefs, and a new one I'm eager to try when I can find a spot for its height: **Spice Islands**, three to four feet tall.

Another approach, which simplifies many decisions: treat rosemary as an annual. Each spring, as soon as you find them, buy a beautiful new transplant, enjoy it till cold weather threatens, and then be prepared to say goodbye. Next spring, buy a new one, maybe some variety you haven't tried before, and continue your adventure.

TARRAGON

I try to avoid saying this or that plant is my favorite type of [whatever]; it strikes me as somehow unseemly, since I write about all of them and should care equally. But sometimes I just can't help it. Tarragon is one of my two all-time-favorite, can't-imagine-life-without-it herbs. (The other is basil.) If you've ever cooked with dried tarragon and been disappointed with the taste, it's not your fault—the dried version of this herb is definitely inferior. (I'd like to use another word, one that rhymes with *waffle*). To see what all the fuss is about, you really need to taste tarragon when it's fresh. And that's why I'm so excited to introduce you.

Tarragon Basics

- Start from transplants. (You have no choice.)
- This is an herbaceous perennial.
- It loves full sun and hot weather.
- Go easy on watering.

Look over that "basics" list again—this is a true Mediterranean plant, one that thrives in hot, sunny weather; is completely at home in sandy, rocky soil; doesn't want us to fuss over it with water and fertilizer regimens; and rewards our benign neglect with a bounty of small, well-behaved, incredibly delicious leaves.

PLANTING, GROWING, AND ENJOYING YOUR TARRAGON

You'll start your tarragon adventure with a transplant, because true tarragon doesn't come any other way. (More on this shortly.) Tarragon is particularly sensitive to, and easily damaged by, waterlogged and compacted soil. So before you put your plant into its new home, it's a good idea to first mix in some sand to your basic potting soil to make sure it drains well and also to make this Mediterranean native feel right at home.

Maintenance. Water the new plant well, to help it get settled in, and thereafter water only when the soil is really dry. I'm going to say this again: the worst thing you can do is oversaturate the soil. Some people like to add a bit of fish fertilizer or liquid seaweed extract a few times during the growing season; others think their plants do just fine without it. I agree with both—meaning I do it if I think of it and don't worry if I don't.

Find the container a spot in the sun, preferably at least six hours a day, and try your best to be a little patient. As soon as the young plant is well established, with the stems showing new green growth at the tips, which may take a few weeks, you can begin harvesting. Snip off the new top growth, leaving at least one third of the existing branch. Cut only as much as you need for today; you want it completely fresh. Tarragon is a cut-and-come-again plant (see page 94), and new growth will sprout from the spot where you cut. Keep doing this all summer long to keep the plant as compact and bushy as possible. (It will never be perfectly tidy.)

In winter, your tarragon plant will die down to the soil. Don't worry—it isn't actually dead, merely dormant. This is what *herbaceous perennial* means: the aboveground part dies back, but the roots are healthy and busy storing up energy for the next season's growth. Going through this cold period is critical for continued good health, so if you live in areas that never see winter, you may have to consider tarragon an annual and start over with a new transplant next spring. (Or grow the alternative known as *Mexican tarragon*; see Varieties on the next page.) By the same token, remember the warm-climate ancestry of this plant; if your winters are severe, be ready to move the tarragon container into a protected area where it can still go through that necessary period of chilling but not actually freeze.

In its container, tarragon spreads by fat underground roots called *rhizomes*. In just a few years, it will develop a snarly tangle that can literally strangle the plant. So every three or four years you'll need to divide it. I mean actually cut it in half, either by slicing down through the center of the plant, through the soil, and into the root mass, or by digging up the entire plant and spreading it out on a work table so you can see what you're doing.

Remember that time back in chapter 3 I told you to add a serrated kitchen knife to your tool kit? This is why; you'll use it like a saw to cut through that dense snarl. Now you have two separate plants. Untangle the roots as best you can; cut away and remove any obviously old and shriveled-up rhizomes, and trim the healthy ones back to about two inches. Replant both, or pot up one as a lovely gift for your favorite cook. Do this in the early spring.

A variation on this task will give you fresh tarragon in the winter, and that's a very good thing. Divide your plant sometime in December, after it has died down. Then give one of the pieces a serious haircut, down to about two inches long, trim the roots to a tidy clump, and pot it up into a small container for your sunniest indoor windowsill. Enjoy fresh tarragon through the winter, then plant this treasure back into your container garden in the spring. Two advantages here: you have fresh tarragon on demand all winter long, plus you have just created some insurance in case your remaining outdoor plant doesn't fare as well as you hoped.

Cooking. Cooking with tarragon is pure joy. It has a delicate, haunting flavor that beautifully enriches dishes featuring chicken, fish, eggs, vegetables, along with classic sauces—and is next to impossible to describe with words. It is often compared to anise, which is the flavoring of licorice, and that makes me shake my head because I do not like licorice even one little bit and yet I deep-down-to-my-toes love tarragon.

What we can say, I think, is that tarragon is a subtle, elegant herb and works best with light dishes (meaning the opposite of hearty or robust). It does wonders with egg dishes like frittata or quiche, with tender fish or shellfish, and it is a longtime classic with chicken. It's no surprise, really, that tarragon is often identified with French cuisine, but don't let that stop you. To get the most from this delicate herb, add it near the end of the cooking process. And I probably don't need to remind you, but I will anyway: use fresh leaves. Not something you cut several days ago, and definitely not that dried stuff in the little jar.

Varieties

There's really only one: French tarragon. It does not produce seeds, so the only way to propagate it is through cuttings. If you see seed packets labeled *tarragon*, that is the plant popularly known as Russian tarragon, and it is decidedly unpleasant, harsh, and bitter. French tarragon is the only true tarragon, and it's available only as transplants.

There is, however, one possible substitute, especially valuable if you love tarragon but live in very warm climates, so warm they have no real winter, which tarragon needs. Find seeds or transplants of the plant popularly known as Mexican tarragon. It is actually a type of marigold (*Tagetes lucida*), and it does produce little marigold flowers, but the leaves are the real point, with their respectable similarity to true tarragon flavor. (Sometimes this plant is called *Mexican marigold mint*; common names can drive us all crazy.)

THYME

Of all the lovely things we could say about this sweet herb, here's one more: it makes for great puns. I've seen the word used in a million punny ways—in slogans, restaurant names, garden-article headlines, messages in greeting cards—and it makes me smile every single thyme.

Maybe that's because I'm so in love with these adorable plants, with their tiny leaves and Tinker Bell flowers, and their astonishing array of flavors. Or maybe I'm just easily amused. Either way, this versatile, easy to grow, endlessly surprising little herb earns its place in your garden many thymes over. (I'll stop now.)

Thyme Basics

- This is a perennial.
- It prefers full sun.
- Start with transplants.
- It is distinguished by its small leaves.
- A dry-land native, this plant is easily damaged by overwatering.
- It tolerates cool temperatures.

GROWING, HARVESTING, AND ENJOYING THYME

It is possible to grow thymes from seed, but I believe you'll be happier if you spend time wandering through the herb aisles of a large nursery, exploring the many delightful transplants. From seeds you get a hundred of the same thing; with transplants you get to try many different types, one of each. Just make sure you're wandering through the *herb* section (see Varieties, on the next page).

This is another of those wonderful Mediterranean natives, and you know what that means: lots of sun, hot weather, and soil that drains quickly and smoothly so that the roots never sit in waterlogged soil. Working in some sand to the container soil before planting will help with drainage. This extra step is very much worth the trouble; too-wet soil is thyme's biggest problem. Go easy (or not at all, some say) with fertilizing.

Maintenance. One nice feature of thymes is that, in relative terms, they are hardier than many other Mediterranean types and will better withstand cool weather. Even so, during a prolonged period of bitter cold, they will need protection—such as a heavy layer of mulch, a move to a more protected space, or maybe the plastic-milk-jug cover you created in chapter 3.

Harvesting. Harvesting is easy; snip off one or more stems about one third or halfway down from the tip, wash and dry thoroughly, and add them to your dish as the recipe (or logic) dictates. Entire sprigs are easier because the leaves are so very small, but if you want the separate bits, it's easy to slide your fingers down the stem and free the little leaves. Now you're ready to fold them into marinades for roast meat, cream sauce for chicken pot pie or steamed vegetables, or herbed mayonnaise for pasta salad or sweet potato salad. Or sprinkle onto roasted potatoes or carrots, or use as a garnish to add a bright green flourish to soups both white and red.

Long-term care. Over time, you will discover that in your garden the lower part of the plant stems is becoming tough and woody, and eventually the center area will be threadbare—I mean leafbare. You can control this to a degree by frequent clipping of the stems, which will force new growth at the cut line, but ultimately you won't be able to disguise the leafbare center. If it bothers you, thank the plant for its service and retire it, and go buy another one or two or ten.

Varieties

When you're at that nursery, soaking in all the many possibilities of thyme, make sure you're in the herb section and not the landscaping area, if the nursery is organized that way. Many thyme varieties are used in large gardens as ornamentals. Yes they are edible, in the sense that they won't harm you, but they do not have the rich flavors that you want in your cooking. Don't worry; you'll have plenty to choose from in that latter group. Including, I have no doubt, some that don't even exist today; new ones show up all the ~~thyme~~ time.

Let's stroll through the nursery. The first thyme we see is the one known as **common thyme**, *common* meaning basic, standard, the original from which others have been developed. The scientific name is *Thymus vulgaris*, where *vulgar* is used in its original Old English meaning as common, familiar. If you find a transplant labeled just plain *thyme*, it may or may not be this one; check the tag for the Latin name. Common thyme, also known as **English thyme** and **garden thyme**, has been popular with cooks for centuries, and still is considered the classic thyme taste. Two similar varieties, which vary only slightly in leaf shape and flavor, round out this grouping of basics. **French thyme**, a chef's favorite, has narrower leaves than English thyme and is generally considered to have a milder, more subtle taste. It is sometimes called **summer thyme**, reflecting the fact that it does not do well in cold weather. **German thyme**, with broader leaves and a flavor described as richer, stronger, more robust, is also known as **winter thyme**, because it handles cold weather better than the other two.

Moving beyond this group of three generically named standards, we find varieties that incorporate a second flavor component. In this double-layered flavor group, the most popular and widely available are the citrus-flavored thymes, where lemon predominates. In my mind, lemon and thyme go together perfectly, and that combination brightens the taste of so many dishes that I can't imagine not having at least one on hand. Maybe we'll find the one named simply **Lemon thyme**, with pink flowers and dark green leaves on upright stems about six inches tall. A variant of that is **Silver Queen thyme**, also lemony but with variegated leaves edged in silver tones and with pale-lavender flowers. Even more dramatic is the foliage of **Golden Lemon thyme**, bright yellow to match the bright taste. **Lime thyme** has light lime-green leaves with a very recognizable lime flavor. And rounding out this citrus group is **Orange thyme**, with leaves of a true orange flavor on four-inch stems that also bear pastel-pink flowers.

We might also like to look into a group that meld a distinctive spice flavor with the familiar herbal taste of thyme. One is **Caraway thyme**, which really does taste like both caraway and thyme, and which grows into a dainty mound about four inches high with sweet pink flowers. Also tempting is **Oregano-scented thyme**, a much taller upright (as much as ten inches) that pops with the taste of oregano and practically begs you to make fresh tomato pasta sauce.

CHAPTER 7

The Good Stuff: Edible Flowers

Flowers in a vegetable garden? Yes. A container vegetable garden? Yes. A garden so tight on space that every plant has to earn its spot? Yes!

If you're still frowning at the idea, here are some things to think about:

- Thanks to their vivid colors and fragrances, flowers attract to your garden the pollinators that your vegetable plants absolutely depend on for healthy production.
- Using flowers in your cooking instantly and painlessly turns ordinary into extraordinary. There is no easier way to create culinary magic.
- Practically everyone is happy to help with a gardening task if it involves flowers, and before you know it, you have a convert.
- Nurturing something for the beauty it provides is reason enough.

THE GOLDEN RULE FOR EDIBLE FLOWERS

Do not use any flowers in food, even as a garnish, unless you know *for certain* that they are safe to eat.

And teach the children in your life to be careful too. Make sure they understand that just because they see you nibbling on a flower doesn't mean that all flowers are OK.

We are, of course, talking about *edible* flowers. Meaning the blossoms that you know *for sure* have been proven safe to eat.

First, Let's Talk About Safety

The basic rule is quite simple: don't eat any flower unless you know for a fact it's safe. But little blips along the way can complicate that simple guideline. Here, in no particular order, are some things to keep in mind.

- Start with reliable information. The internet has given us access to an incredible amount of material, not all of which should be taken as gospel. For solid information that is both university-level accurate and also targeted to your area, I can't think of a better, more accessible source than your county extension office.
- Just because one particular flower is edible, others that are closely related may not be. On a similar note, it is possible that while the flower is edible, other parts of the same plant (leaf, perhaps) may not be, and vice versa.
- Children copy what they see adults do. If they see you eat *this* plant, they may assume it's OK to eat *that* one. Make sure they understand how serious it can be to eat the wrong thing—and for those too young to discriminate, not to eat *anything* from a plant without a grown-up nearby. Finally, always keep the phone number of your local poison control office handy.
- Do not cook with flowers you bought from a florist, even if it's on the edible list. They were almost certainly grown in a commercial greenhouse, under conditions you cannot possibly know. Especially if you're considering a flower outside its natural season (roses in February, say), there is a risk that the plants were cultivated with substances you do not want to put in your mouth. And even though I probably don't need to say this, don't use any of those things on the flowers in your own garden either.
- If someone you're cooking for has a tendency toward allergies, explain about the flowers. They may want to try a small sample first or skip the flowers altogether.

For this chapter, I selected four popular flowers that I believe will be rewarding in your garden and in your kitchen—well loved, readily available, offering a variety of tastes, and in bloom at different times of the year, giving you a long-running flower show. But these four are not the only edible flowers, not by a long shot. They are simply the ones I believe will give you the greatest joy with the least angst. Also, don't forget that the flowers of all herb plants are edible. They tend to be small and modest in appearance, but they all carry hints of their plant's herbal flavor, and I believe you will find many ways to incorporate their dainty charm into your cooking.

Flower Prep, Step-by-Step

This part is easy. It's mostly common sense.

1. Snip off the flowers along with a good length of stem from the plant.
2. Holding by the stem, swish the blossoms facedown through a bowl of water, to dislodge any dirt, dust, or tiny stowaways.
3. If dinner is still hours away, put the stems into a "holding" vase, just like a bouquet.
4. When you're ready to use, clip away the stems and do a final inspection. Spread the blossoms (or petals, if you are using them individually) in a single layer onto a plate covered with paper towels or plastic wrap, and set it aside in a cool spot until the very minute you need them.

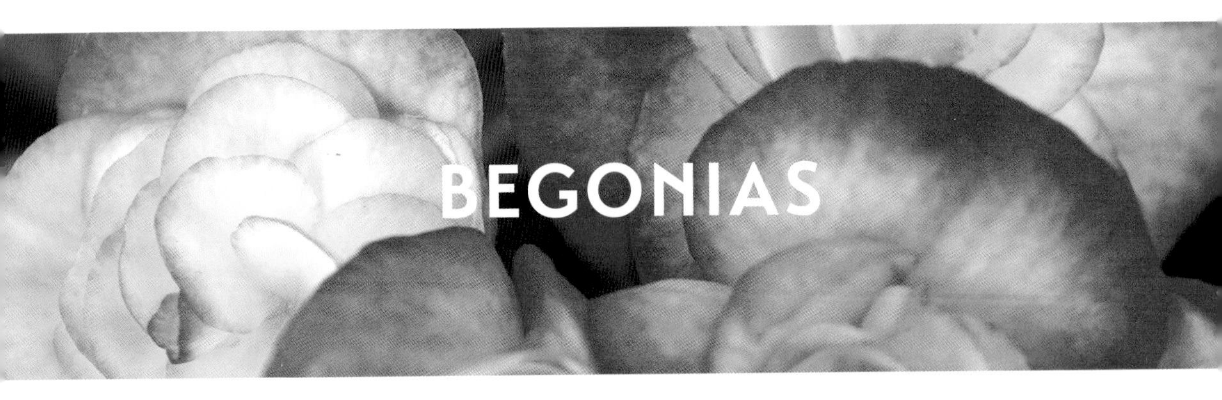

BEGONIAS

It's possible you might not be familiar with begonias, and if so I'm a little bit jealous, thinking about how much fun you'll have meeting this new beauty for the first time. It's extra special in several ways: absolutely drop-dead gorgeous colors, a delightful lemony taste you don't expect from flower petals, and a garden rarity—a flowering plant that doesn't need full sun. In fact, it actually prefers light shade.

Important note: we're talking about the group known as *tuberous begonias*, meaning that they grow from underground tubers. This is *not* the same as the small bedding plant called *wax begonias*. Nor is it the type called *angelwing* (for the shape of its foliage); that one is usually considered a houseplant.

Begonia Basics

- Start with tubers or transplants.
- This is a perennial, although not long-lived.
- It needs part shade.
- Careful fertilizing program equals more flowers.

GETTING TO KNOW TUBEROUS BEGONIAS

A tuber is one kind of underground storage system from which new plants grow, something like a squashed or elongated bulb. For the edible begonias described here, the tuber is the distinguishing characteristic; that's how you know you're getting the kind of begonia I want you to have. But once it's planted (see next page), you can put the tuber out of your mind, at least until wintertime (and maybe even longer, depending on your winter climate). We have more important things to focus on. First and foremost, the flowers.

These flowers are a total knockout. They come in a broad range of colors, from deeply saturated hues to soft pastels to gentle shades of ivory and even to two-toned petals with a delicate contrasting edge, and they all have an amazing luminous glow that always makes me think of stained-glass windows. The individual flowers stay on the plant for a satisfyingly long time, and the bloom season is long, from early summer to late fall. Here in western Oregon, I often have flowers at Thanksgiving.

One trait of the flowers that is always fun to explain to visitors is their sex. Yes, there are male and female flowers. The males are quite spectacular—large and full of petals. The females are much more modest, with just one flat layer of petals. The blossoms often appear together on one stem in groups of three: one princely male surrounded by two quieter females. Mother Nature's gentle joke.

The petals have a tart citrusy taste that is very much like fresh lemon. The combination of lemon flavor and bright color makes them almost endlessly versatile for adding a splash of both to your cooking.

The leaves are rather large, triangular shaped, dark green to bronzy green in color, and slightly fuzzy. We don't need to spend too much time on them, except to make sure they aren't blocking the full development of the flower buds that may be hiding underneath. If you find that problem, carefully snip off the entire leaf from its branch. Coincidentally, this also helps with another possible problem: because of their size, the leaves have been known to put so much weight on the very tender main stems that whole branches break off, taking the flower buds with them. This danger is especially high when the leaves get wet, so be careful with your watering.

PLANTING AND NURTURING YOUR BEGONIAS

Early in spring, you can find tubers at online nurseries or retail garden centers. The tuber comes packaged with printed instructions and (usually) a color photo of the bloom. Most mail-order catalogs will also feature gorgeous photos to help you zero in on the colors you want (or drive you crazy trying to decide).

A mature plant can be twelve to eighteen or more inches tall and wide but does not need more than six inches of soil depth. The tuber itself looks rather like a tulip bulb that

Bigfoot stepped on, leaving a small indentation in the center. Plant it directly into your container, with the indentation facing up, after the spring frost date (review in chapter 2), and watch for pointy shoots in about six weeks. In the meantime, keep the soil lightly moist and be careful not to plant anything on top of the tuber. (This might be a good time for the radish tip on page 130.)

Another approach, and simpler, is to wait a few weeks until transplants are available at the nursery. One big advantage is that they may be showing their flower so you can tell exactly what color you'll be getting. One possible disadvantage: you can't always tell from the plant tag whether this is an upright or trailing type (described on the next page). Unless you have your heart set on creating a hanging basket, this may not matter to you.

Shade lover. This might seem strange to veteran flower gardeners, but keep this plant away from direct sun. Partial shade is best, but they will also do fine in full shade. This is very good news for container gardeners who love growing flowers but often struggle with limited sunlight.

Maintenance. Through the growing season, your begonias will need you to pay attention to both water and fertilizer. Don't let the soil dry out completely; use the finger test often. But when you do add water, direct your hose or watering can to the edges of the container. Try not to water the plant itself, especially the leaves; as we know, when waterlogged they can break the stems from the extra weight. And to keep those beautiful flowers coming, give your plant a weekly dose of complete fertilizer that is high in phosphorus.

Winter care. Your task at the end of the blooming season varies depending on your winter weather patterns. In very warm areas, you don't need to do anything. The plant may go dormant and drop leaves, but the tuber is still healthy and will regrow come spring. If your winters are relatively mild, without long periods of near-freezing temperatures, you can usually manage by moving the container into a more sheltered area or wrapping it in some insulation (Bubble Wrap, say). But in severe winters your best bet is to dig up the tubers and save them for next year.

The process is not complicated. Dig up the plants, trim away any remains of foliage, dust off the tubers, and let them dry thoroughly. (If you started with tubers, think back to what they looked like when you opened the package; you're trying to recreate that condition.) Then place them into a small bucket filled with sand or peat moss, and set the whole thing aside until next spring. If you use paper bags instead of the bucket, you can write on the side what colors are hiding inside. Or you could just start over next year with brand-new plants; that works too.

Varieties

I'm not going to suggest specific varieties here for one big reason: most of the specific cultivar names for begonias merely identify the shape and color of the flowers, and you can determine those qualities much more reliably with your own eyes by looking at transplants in the garden center or studying the catalogs of mail-order nurseries. Instead, I'm going to point you toward a large category called **Nonstops**, which exactly fulfill the promise of the name: they bloom nonstop from early summer through fall. You can identify Nonstops by the plant tags in the nursery or the catalog descriptions, and as long as you attend to watering and fertilizing, I promise you will not be disappointed.

Within that category are two different types, distinguished by the way they grow: on *upright stems*, eventually forming into the shape of a petite shrub, or *creeping branches* that will trail over the edges of the container. But in reality the difference is relative. Even the upright stems, laden with flowers and large leaves, tend to lean downward through simple gravity, giving you a mildly cascading appearance even if you didn't plan for it. And it looks terrific, so don't worry. But if you are designing a hanging container, you definitely want the cascading type, so pay attention when ordering or browsing; you'll still find many incredible colors.

Cooking with Begonia Flowers

Remember that the basic taste is lemonlike, and many possibilities will pop into your head. Then think about color and ask yourself, *Where could I use the combination of lemon plus that beautiful color?* I'll start you off with a few suggestions.

- Stack individual petals high and slice into bright ribbons for garlic mashed potatoes or shrimp scampi. For example: yellow ribbons on watermelon salad or grilled tomatoes.
- Slice the stack in one direction and then the other, for tiny lemon-flavored confetti to sprinkle on cream soups or roasted vegetables. Add pink confetti on cupcakes with cream-cheese frosting. Blend red confetti into cooked rice, for your special lemon rice in which the bits of "lemon" are not yellow but bright red.
- And here's a summertime dessert idea that promotes togetherness and giggles. In the center of a large platter, place several small bowls with different types of fruit-flavored yogurt and surround them with begonia petals in different colors (whole, not chopped). The petals, which have a naturally crisp texture, serve as dippers for the yogurt. Try as many tart-sweet combinations as possible.

DIANTHUS

So far, I've resisted the urge to explain scientific names (even though I think they're fun to know about), but I'm going to make an exception here because otherwise we'll land in a taxonomic briar patch. *Dianthus* (in italics) is the name of a genus (large group) of plants that comprises many species; it also includes one species that is called dianthus (no italics), for no logical reason other than that's the way popular usage evolved over time. The no-italic dianthus is what we're talking about here.

It has another name too—*pinks*—which has nothing to do with color but instead refers to the small notches on the edges of the petals, as if they had been trimmed with a seamstress's pinking shears. Or for all I know, that form of edging is named after the flower petals; the history of word usage is a fascinating tangle. In any case, just keep in mind that the flowers of pinks can be white, red, lavender, or a million shades of pink.

Dianthus Basics

- Start with transplants (recommended) or seeds.
- These are perennials but relatively short-lived.
- They bloom in spring and early summer.
- They want full sun, regular watering, and a bit of fertilizer now and then.

GETTING TO KNOW YOUR DIANTHUS

In your garden, these are small, tidy plants with dainty flowers about the size of a quarter. The feathery foliage forms a loose mound about six to eight inches high and wide, and the flowers appear at the top of stalks that are as much as six inches higher.

Several different plants massed together in a container make a lovely display from spring through summer.

These are perennials but are not especially long-lived. Like most flowering plants, they are happiest in full sun and will keep madly producing flowers with the help of a little fertilizer, regular watering, and regular trimming. The blossoms have a mild floral taste, but their contribution to food is mostly color—except for one type, known as *clove pinks*, which carries the taste and the heady fragrance of actual cloves.

PLANTING AND NURTURING DIANTHUS

This sweet little plant is nicely trouble-free. If you want to practice your seed-cultivation skills, start seeds indoors about a month ahead of your spring frost date, then transfer to your outdoor containers once freezing temperatures have passed. Just remember that briar patch of *Dianthus*/dianthus names, and be sure you get the seeds that will produce the plant you have in mind.

You may prefer to buy transplants. Not only is this simpler overall, it's a way for you to see the actual color of the flower, which may be important if you are, for instance, planning a mixed container with a color theme.

These little beauties do most of their blooming in the coolish days of spring and early summer, but like all flowering plants respond well to a modest fertilizing regimen with a complete fertilizer high in phosphorus. And if you keep the spent blossoms clipped off, the plants often put on a second flurry of flowers in late summer. These plants are perennial but tend to be rather short-lived—three, four, or five years. And they will be glad for a bit of protection in very cold weather; see chapter 3 for ideas.

Varieties

Since the main trait I suspect you will be interested in is the color of the blossoms, I'm not going to suggest specific varieties. Your eyes will make the decision for you. The one thing I hope you will consider is including at least one **clove pink**. The fragrance and the taste are so astonishingly close to the familiar spice. It always surprises people who try one for the first time.

Cooking with Dianthus

You have a chore first: remove that white "heel" at the base of the petals. It's unpleasantly bitter. The easiest way to do this is to grasp with one hand the entire flower head, while it is still on the plant, and with the other hand cut the flower head away from the stem, just above the white base. Your first hand now holds individual petals sans bitter base. Snip off and discard the rest of the decapitated flower stalk.

Now, imagine what culinary creations would benefit from a sprinkle of those pretty flower petals—especially if you're using tasty clove pinks, which give both color *and* the taste of cloves. Picture baked goods and desserts, flavored beverages, and classics like baked ham with cloves. Can you steal ideas from any of them? A few suggestions:

- Hot beverages for wintertime: lemonade, black or herbal tea, hot toddy, mulled wine, Irish coffee, hot chocolate with whipped cream topped with flower petals.
- Fruit dishes: Add to the liquid for poached pears. Work into the top crust for berry pies. To accompany Sunday morning pancakes, sauté apple slices in butter enriched with petals.
- Savory sides and entrées: Sweet potatoes, peeled and roasted whole, sliced thick horizontally, then arranged on a large serving dish with pats of butter onto which you have scattered dianthus petals. Open-face tea sandwiches on date nut bread, spread with a mixture of petals and softened cream cheese. Roast pork tenderloin, sliced into individual servings, topped with peach chutney and a sprinkle of petals.

I was born in 1943 in Montpelier, Idaho. It wasn't until after World War II that I first became aware of the term *Victory Garden*, but I well remember my grandparents' garden from those days, even though I didn't know the name and they probably didn't call it that. To them, it was just *the garden*.

My grandpa, a conductor on the Union Pacific railroad, was responsible for the vegetables. He grew just about every vegetable you could think of, but my favorites were the carrots, which I was allowed to dig and eat right there in the garden, with the dirt rubbed off on my Levi's. They also had chickens and a cow that I never got the hang of milking. My grandmother took care of the flowers, peonies and lilacs that were always in bloom in time to be taken to the cemetery on Decoration Day [what we now call Memorial Day].

I'm sure I inherited gardening genes from that side of the family, and to this day my two favorite flowers are peonies and lilacs. Wherever I have lived since then, I always tried to have a garden whenever possible, some large and some not so large. Today I'm trying to garden in pots, in a very small area off our kitchen. So far, some things have been very successful, some were a complete bust. The experiment continues.

Nyel Stevens
Oysterville, Washington

NASTURTIUMS

Nasturtiums bring a riot of color to your garden all summer long and into the fall. They also bring our friends the pollinators—bees, butterflies, and hummingbirds. The flowers have a unique structural feature at the base of the petals: a long, thin funnel filled with nectar. Technically it's known as a *spur* but most folks call it a tail, because that's what it looks like. Look at a flower in profile and it will be immediately obvious why hummingbirds love this plant.

Gardeners love them for many reasons, not the least of which is how easy they are. They don't require any special soil, can tolerate a minimalist water regimen, and actually perform better if not fertilized at all. There are two main types, defined by how they grow: into short, compact mounds or long, trailing vines. More information is in the following pages; for now I just want you to be aware of this difference here at the beginning because it affects your early steps.

Both the flowers and the leaves are edible, and both have a peppery taste that reminds some people of watercress. You may already know them; it seems to me that this was one of the first edible flowers to find its way to American tables. But if your acquaintance has been limited to the very familiar orange blossoms, you have a wide new world to explore.

Nasturtium Basics

- Nasturtiums are annual plants.
- Start with seeds.
- They prefer full sun but will tolerate partial.

- They forgive occasional dry soil.
- Minimal fertilization equals maximum flowers.
- Remember the two different types: trailing and mounding.

PLANTING AND GROWING TIPS

Try to find a spot in full sun. If necessary, nasturtiums will tolerate partial shade but will produce fewer flowers. They are not fussy about soil; whatever you use as your basic potting soil will be fine. Nasturtiums do not transplant well, so you should always start with seeds. Luckily, this is not hard.

Before planting, soak the seeds overnight to soften the very hard outer covering. Plant them three to four inches apart and about half an inch deep. Be sure the seeds are fully covered; they need complete dark to germinate, which will happen in about seven to ten days. If you are growing one of the mounding bush types, thin the little seedlings to about ten inches apart; they need room to spread. The trailers can be closer together.

PLANT HEALTH

Keeping everything healthy is, in many respects, an exercise in benign neglect. Heavy watering and religious fertilization will give you massive plants with lots of leaves and very few flowers—the opposite of what you want. In fact, your plant will do just fine with no additional fertilizer and only minimal watering. Add water in dry spells; otherwise just check periodically. (Therefore, don't plant nasturtiums into a container that will also be home to plants that *do* need intense watering and fertilization.)

You won't encounter major problems with plant diseases or harmful insects, except for one little devil—aphids. For some reason, they just love nasturtiums. In fact, people with large gardens often plant nasturtiums off to the side deliberately to draw aphids away from their crops. Container gardeners have it easier because our space is so much smaller and individual plants get our tender attention every day. Still, whenever you gather flowers and leaves for your dinner, be sure to check the undersides of the leaves. That's where these guys hang out.

HARVESTING

This is a simple matter of snipping off flowers and leaves with your clippers or kitchen scissors. And of course the more flowers you pick, the more new ones you will have. Your plants will keep producing into the early days of fall, at which point you might want to leave a few flowers aboard to set seeds. Dust them off, spread out to dry well, and set aside for next year. If you miss some, you may well have some volunteers pop up next spring.

Varieties

The first thing we need to focus on at this point is something hinted at earlier: the two very different types of nasturtiums, identified by their growth habits, plus a third, less precise one somewhere in the middle: trailing, mounding, and semitrailing. It's critical to container gardeners because it firmly dictates where each type can be grown successfully. As is often the case, the terminology is something of a jumble. *Mounding* equals *bush* equals *compact* equals *dwarf. Semitrailing* or *semibush* or *semidwarf* mean the same thing—in between mounding and trailing; which terms you use depends on which end of the spectrum you start from. The only one that's not complicated is *trailing*, which has limited usage in container gardens.

GROWTH HABITS

Trailing nasturtiums. These are distinguished by their long stems; they actually develop into vines that can be as much as six or eight or even ten feet long. If you accidentally plant this type in a standard container, they will grow down the sides and all along the floor in every direction, and pretty soon you have a mess on your hands. Your only option is to plant them up high.

If there's a tall fence or wall where you live and you can affix a planter securely to the top, you can create a dramatic living tapestry as the vines grow downward and cover the entire space. Planting them at the base of a fence is also possible, especially something with narrow vertical pieces, although the vines will need some help from you as they don't have natural "grabbers" like peas. I can also picture them growing onto a large trellis, perhaps attached to the side of your garage; here again, you'll have to help the vines grab on. You could plant them on a balcony rail and let them dangle naturally, if you're on good terms with the neighbors below. And if you have a location where a hanging container can be positioned very high, that will work also; the standard hanging baskets may not be high enough.

Mounding nasturtiums. As the name suggests, these nasturtiums grow into a mound; generally, the mature plants reach a height of ten to twelve inches and similar width. They are also known as *dwarf* or *bush* nasturtiums; all those terms mean the same thing. These are the ones most appropriate for container gardens because, compared to the other two, they maintain a compact, bushy shape. This is also where you will find the greatest selection of stunning colors.

Semitrailing nasturtiums. Semitrailing types (or semibush, depending on your point of view) occupy a middle ground between the two: the stems are actually short vines that dangle but not as long, generally two to three feet, and the overall shape tends toward a kind of loose mound. The dividing line is imprecise and not universally agreed upon. Depending on the source, some varieties with twelve-inch stems may be called a *bush* or a *semibush*. This may be another case where we care less about what something is called than about how it behaves in the garden. So, my conclusion: varieties in this category, whatever we call it, are the best for hanging baskets.

Keep in mind that some of the longtime favorite varieties are available as both types—mounding and trailing—so pay attention when buying.

FLOWER COLOR

The second trait to consider—and a whole lot more fun—is color. If you've ever seen a nasturtium plant in bloom, there's a good chance you saw orange flowers; that's by far the most common color. But, oh my goodness, there is so much more! The nasturtium world features so many new varieties in sumptuous colors, all so gorgeous you'll have about a million possibilities to choose from. And that's a problem in itself, isn't it? I'll try to help with some specific suggestions, but your very best bet is to hang out with the seed racks at the best nursery you know, and look at the individual photos. The packet itself will tell you exactly how large the mature plant gets—if it doesn't, put that one back—and very often will also indicate whether it is a mounding or trailing type. That plus your reaction to the color is really all you need.

One of the first things you'll notice is that many seed houses have created special mixes—seeds of closely related cultivars in several different colors together in one packet, for those who like the casual look of a happy jumble. The word *mix* will be part of the name, like "Jewel Mix." If you lean more toward the single-color style, often the individuals in those mixes are also available separately, like "Jewel Cherry Rose."

Some suggestions to get you started:

Mounding types. **Alaska** is notable for variegated leaves splashed with white, familiar orange blossoms, and a mature height of twelve to fourteen inches; it is often sold as a mix. The **dwarf form of Jewel** (see the following page) grows to ten to twelve inches.

Whirlybird comes in many hues of yellow, orange, salmon-pink, and cream; grows to ten inches; and is often sold as a mix. **Ladybird**, best known for its unusual bicolor flowers in yellow, pink, or cream with red splashes, reaches eight to ten inches and is often sold as a mix. **Vesuvius**, with its delicate salmon-colored blooms, reaches twelve inches.

Semitrailing types. **Empress of India**, with rich scarlet flowers against blue-green leaves, has a mounding habit but with vines that trail one to two feet. **Jewel**, beloved for its brilliant colors, grows to eighteen to twenty-four inches. Check out the **Gleam** series, All-America winners dating to the early 1930s: **Golden Gleam** (AAS 1933) is a beautiful pure yellow, **Scarlet** a gorgeous red, and **Glorious Gleam** (AAS 1935) is a mix of several colors; all reach a size of two to three feet. **Firebird**, with orange and yellow flowers above variegated leaves, grows one to two feet. **Orchid Flame**, a mounding habit with long vines two to three feet, is noteworthy for its unique flowers, which start out a golden yellow with red splashes, then gradually change color to fully red.

Trailing types. **Jewel of Africa** has variegated leaves and red, yellow, pink, or cream flowers and grows to four to six feet. **Apricot Twist**, apricot-orange flowers with red streaks, is a little smaller, at three to four feet. **Purple Emperor**, soft lavender-purple flowers with yellow throats, is smaller still, at two feet. **Amazon Jewel**, from Renee's Garden, with brilliant flowers in a range of colors, reaches four to six feet. **Moonlight**, noteworthy for the soft-yellow tones of the flowers, likewise runs four to six feet. **Spitfire**, from Renee's Garden, features brilliant scarlet flowers and grows to four feet.

And now for a gorgeous new color making garden news. In 1935, All-America Selections named Gleam nasturtiums a winner. It took eighty-four years before they found another worthy nasturtium, and it's a doozy: **Baby Rose**, with a mounding growth habit and delicate blossoms in a knockout deep-rose color. Then in 2020, another winner: **Tip Top Rose**, with mounds about fourteen inches tall and eighteen inches wide, and masses of gorgeous flowers. And from Renee's Garden, **Cherries Jubilee**, a mounding type ten to twelve inches tall, in many shades of rosy red. Any one of these will be a stunning addition to your containers.

Cooking with Nasturtiums

Both the flowers and the leaves are edible, both with a mildly peppery taste that has been described as similar to watercress with a bit of sweetness added. In both flowers and leaves, the taste is more appealing when they are young; older leaves in particular become bitter the longer they stay on the plant.

Generally, we have focused our culinary attention almost exclusively on the flowers, but I want to encourage you to think about the leaves too. They are high in vitamin C, richly colored in shades of green, and a very pretty round shape that looks like a lily pad for fairyland frogs. Here are a few possibilities:

- Include in salads of mixed greens.
- Use with sandwiches in place of, or in addition to, lettuce.
- Line serving dishes or platters with overlapping leaves.
- Use the lightly cupped leaves as tiny shallow bowls, holding an individual serving of a condiment that accentuates your entrée. Present that entrée in the center of a serving platter, surrounded by a generous number of those tiny leaf bowls for people to help themselves. Possibilities: your homemade cranberry mustard for Thanksgiving turkey; herb butter for poached fish; parmesan-garlic yogurt sauce for roasted oysters; crème fraîche blended with orange marmalade for baked ham; ricotta cheese mixed with blue cheese for grilled steaks.

(I know you washed everything carefully first, but remember to double-check the undersides for rogue aphids.)

The flowers, with their bright colors, present more options. The flavor, while still a bit spicy, is more floral than that of the leaves, and they can be used whole or diced into smaller bits. Whole flowers make a dramatic garnish on just about anything. And if you gently pull apart the individual petals, you have tasty, colorful tidbits that will add your special touch to many savory dishes. For example:

- Soups, especially in contrasting colors. Mac and cheese. Scalloped potatoes. Spaghetti alfredo with shrimp. Oven-roasted cauliflower.
- Create a fancy pattern on a wheel of brie for the appetizer table.
- Dice into tiny bits and mix into hummus.
- Incorporate into softened cream cheese for bagels.

Can a small backyard ditch hold a story? In my case, absolutely.

One afternoon in 1954, I was helping my father gather tomatoes from a small garden patch in the backyard of our home in West Hartford, Connecticut. At some point I glanced over at the remains of a shallow ditch nearby. My seven-year-old self knew the ditch as the place I used to sail my toy boats after a heavy rainstorm; it had never occurred to me to ask why it was there.

"Victory Gardens," he said. "No garden, no fresh vegetables." It was a long time before I understood the full story of that marvelous neighborhood Victory Garden. Or appreciated the fact that the ditch was all that remained.

During World War II, all Americans were encouraged to plant a Victory Garden wherever they could. Somehow (I never learned exactly how), five families all in a row, including ours, decided to put in a garden, and soil near the back property lines was dug up and used to construct the garden plots. That left a shallow ditch about three feet wide spanning all five backyards. Great for sailing homemade boats and otherwise getting in some kind of trouble.

I doubt that the neighbors, all city folk, knew anything about gardening for food, but Dad did. The entire backyard of his childhood home in Fall River, Massachusetts, was a garden—a necessity for his large family. Grandpa Boyle, an Irish immigrant and a widower, worked in an iron foundry away from home all week, so the seven children took care of the garden full time after their mother died. On my last visit in 1982, that garden was still feeding the entire neighborhood.

Dad was the primary gardener of the five-family Victory Garden on Westbrook Road. And in addition to his civilian job, he served as civil defense warden for the West Hartford area from December 7, 1941, until the summer of 1944, when he was drafted. He was sent to Germany as a mess sergeant with a hospital, but he was always a farmer at heart, and all throughout Europe lived off the land whenever possible. He even hauled two pigs across the battle zones, often ahead of the troops, just in case they ran out of food. Those pigs lived well and were given to a farmer after the war.

Mary (Dee) Ponte
Hertford, North Carolina

PANSIES AND VIOLAS

These delicate beauties, beloved since Shakespeare's day, would be a welcome addition to any garden just for their sweet charm alone. They are small, well-behaved plants that in containers will form soft mounds about six inches high and six to eight inches wide, with the flowers on stalks a few inches taller so that they seem to form a kind of loose carpet over the green leaves, a magic carpet of adorable flowers absolutely no one can resist. A happy bonus: they are also edible, with a tender floral taste to complement their delicious colors.

In this section we are considering pansies and violas together because they are so very closely related; in fact, they look so much alike that violas are sometimes assumed to be just miniature pansies, and that's not far wrong. Both belong to the *Viola* genus, and that name has become the common name for the smaller plant. The most obvious difference is with the flower sizes: pansies are at least twice as large. The two types of plants are very similar in overall size and growth habit, and also in what they need in terms of care. So henceforth, for the sake of simplicity, I'll just use the term *pansies* to mean both.

Container gardeners love pansies for several reasons. They bloom early, in the first days of spring when we are all yearning for flowers. And they will, with a little encouragement from us, rebloom in the fall, when most other flowering plants are closing up shop. The plants are small and tidy, so we get maximum use out of our limited spaces. And for parents and grandparents, it's a wonderful way to introduce children to gardening; they are invariably enchanted by the kitten-face flowers.

Pansy Basics

- These are cool-weather plants (spring and fall).
- Start with transplants.

- Grow as annuals.
- They prefer full sun (ideal) or partial shade.

SUCCESS WITH PANSIES

First, let's talk about seasons; it's important. These are cool-season plants that flourish in spring *and* fall but absolutely collapse in hot weather. So you can have spring pansies or fall pansies or both, as long as you plan accordingly. This is mostly a matter of integrating them into your seasonal plans.

Planning and design. In spring, if you're growing other cool-season plants, the pansies will fit right in just about anywhere, since they take up so little relative space. Or, if you intend to add warm-season plants later on, leave the center of the container clear, and position the pansies as a colorful border. Once the pansies fizzle out in the heat, they will need to be dug out and discarded, opening up space for the center plant to expand or for something else to be added.

For fall, you need to find a spot that has room for the new pansies, a spot you're not going to need again until next summer. Here's why: except for the very coldest parts of the country, fall pansies frequently winter over, meaning that they stop flowering but are still very much alive. Next spring they will burst back into life, so they need to stay undisturbed in the meantime.

A lovely idea for both seasons: a mass of pansies in a single container by themselves. This can be either all one color, which is dramatic and elegant; or a mix of colors, which is informal and playful. I personally like them in round, shallow containers in complementary colors; pansies have shallow roots and don't need more than six inches of soil depth. And position the plants very close together, with about two to three inches separating the outermost edges; this is what will create a dramatic color explosion.

Planting and maintenance. It is, of course, possible to start pansies from seeds, but it is such a lengthy process that I think you'll be much happier with transplants from the nursery. As a nice bonus, you'll be able to see the colors of the blossoms, which is really the point. Another nice bonus: it removes uncertainty about the exact best time for planting pansies in your particular climate. When they show up in your local retail nurseries, that's your cue.

Plant the new babies in the spot you have picked out for them. Water well, especially if this is a fall planting and the soils are still summertime-dry. Work a little compost into the planting hole first, if you have some on hand; if not, plan to add a boost of liquid fertilizer after a few days. The final position for your planted container should be where it receives six hours of sun a day; pansies can live with less, but they won't bloom as profusely.

Maintenance is essentially the same in both seasons, and it is not difficult. Pansies prefer consistently moist soil. Not waterlogged, of course, but try not to let them go really dry. Use the finger test frequently, especially in warm spells: if the soil is dry one inch down, it's time to water.

We grow these plants for their flowers, which means they respond well to a steady diet of nitrogen-rich fertilizer. While they are actively blooming, every two weeks give them a dose of complete fertilizer. For fall types, taper off on the fertilizer toward the end of the season.

Varieties

At last count there were something like three hundred different pansies and violas, and more new ones are being developed all the time. They come in the most glorious palette of colors you can imagine and a range of solid colors, double and triple colors, with contrasting streaks, splashes, dots, and painted edges. And since I believe it's the color you are most interested in, the only realistic way to choose is to wander the aisles in a well-stocked nursery and let yourself be seduced by all the offerings. Oh, and some are also fragrant, especially solid yellows.

Cooking with Pansies

Pansies have a mild taste that is a little bit floral and a little bit spicy and, some say, also a little like mint. Truth is, though, it's their astonishing colors that creative cooks are most interested in, and so they are mostly used as garnishes.

The possibilities for new ideas hinge on the question, What exactly are we going to garnish? Cakes and salads are classic, and deservedly so, but maybe we can widen our thinking a bit. Remember that violas are much smaller, small enough to fit into a soup spoon. Also remember that both types of flowers are made up of individual petals, usually five. Sometimes a single petal is just the accent you need. Always try to arrange a color contrast—that's what makes garnishes work.

Here are a few suggestions:

- Green salads garnished with pansies in shades of orange or yellow. This is most effective if the salad itself is mostly greens, without additional vegetables in other colors.
- Wilted spinach salad. Hit with the warm dressing, the spinach leaves become an even brighter green; add the flowers just before serving.
- Fruit salads. Use dark blue or purple flowers for best color contrast.
- Soups, either hot or cold. Best for monochromatic soups that have a smooth texture, such as cream soups. Think in terms of color contrast: blue or purple with cream of tomato, deep orange on cream of asparagus or broccoli, multicolored on potato soups. Add the garnishes just before serving.
- Deviled eggs. Use whole blossoms to garnish the serving platter, individual petals on the eggs.
- Tomato aspic. For a July Fourth picnic, create a salad in red, white, and blue. Individual servings of aspic, covered with a thin layer of spiced-up mayonnaise, and then a blue pansy on top.
- Molded cranberry salad for winter holidays. For each individual serving, add a dollop of mayonnaise or crème fraîche, then top that with one small flower.
- Fancy ice cubes for summertime parties. Fill the trays half full, lay one perfect blossom in each individual cell, freeze just until the flower is firmly held in place, then fill the tray to the top and freeze solid.
- Cupcakes or miniature cheesecakes, with one large pansy covering the top.
- Garnish atop fruit sorbet.
- For kids, ice cream served in small paper cups with a kitten-face pansy on top.
- And of course, unique decorations for cakes. The most beautiful cake I ever saw was covered with royal icing in an ivory shade; all around the sides were large, rich-purple pansy blossoms, exactly the height of the cake and their edges touching, so at first glance it seemed they were holding the cake upright. The brilliance of simplicity made stunning by the regal colors.

A CLOSING THOUGHT

You have probably noticed that we're hearing a lot these days about pollination and how important it is to the world's food supply. At the same time, many gardeners (including some quoted in this book) are organizing their gardens deliberately to provide resources for pollinators. Turns out, they are talking about flowers. So let us close this chapter with a short, very simplified description of pollination and pollinators and one easy way we can all help with this critical need.

In essence, *pollination* is the process by which flowering plants are able to produce viable seeds for reproduction. In the broad landscape of all flowering plants, more than 1,200 are food crops. So anything that interferes with pollination is serious indeed.

It all starts with a substance called *pollen*, produced by the stamen, the male part of the flower.

Pollen gets transferred to the stigma at the top of the pistil, the female part.

The pollen then travels down the slender pistil to the ovary at the bottom and fertilizes it.

Seeds develop in the ovary, surrounded and protected by fleshy material we call plums or cucumber or butternut squash.

If pollen does not get transferred, there will be no seeds for next year.

And that's where pollinators come in. That's the collective name we have given to certain creatures that consider pollen as food and that actively climb around inside a flower looking for supper. The pollen is sticky (which is how it stays in place), so a hungry bumblebee gets some on its legs and accidentally transfers it to the female pistil as it continues to forage. Bingo—fertilization accomplished.

Who are these pollinators? Bees, ants, butterflies, and some birds. How can we help them?

Ah, that's where gardeners come in. A pollinator, continually hunting for the pollen and nectar that sustain it, is drawn to a flower because of its fragrance, or color, or both. And even if that particular flower is not on what we would consider a food plant, the pollinator is now in the immediate neighborhood of other plants that *are* food producers and will welcome them in.

In short, then, growing flowers brings pollinators to your garden. Even if your garden doesn't need them, your neighbor's might, or her neighbor's. And that is how, collectively, the gardeners of the world can help assure a healthy diet for all.

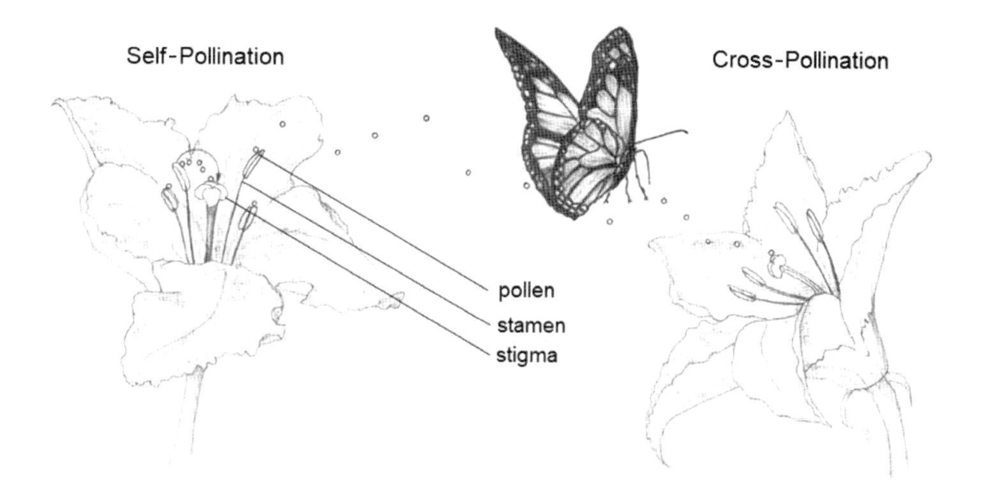

A lovely butterfly, nibbling on pollen, accidentally pollinates both that flower (called *self-pollination*) and another nearby. Both are now able to produce viable seeds for next year.

Community Garden

EPILOGUE

A Ribbon Through Time

Throughout this book, you have read stories from people who vividly remember their family's Victory Gardens; many of them trace their own love of gardening back to those days. In this way, their stories carry the implicit notion of an abstract link between the families who cared for those 1940s gardens and the lives of today's gardeners, who stand with gratitude in their shadows. But I also found a quite wonderful circumstance in which that link is not abstract at all but very real and very tangible. In fact, I found three.

In Washington, DC, in Minneapolis, and in Boston, there are thriving community gardens on the exact piece of property once set aside for a Victory Garden in World War II, and they have been fiercely defended against commercial-development pressures ever since. Think about that for a minute: for eighty uninterrupted years, gardeners in the middle of three of our biggest cities have been able to grow fresh vegetables in original Victory Gardens. I would not be surprised, in fact, to learn there are more, but these are the three I am aware of. For this book, I had the pleasure of talking with six of the modern gardeners, and I'm honored to share their stories.

Cleveland Park Community Garden, Washington, DC

In 1942, as part of the nationwide Victory Garden campaign, the National Park Service designated land in Washington's Cleveland Park neighborhood as a space for sixteen Victory Gardens. Today, Cleveland Park Community Garden, still considered part of the National Park system, consists of more than one hundred plots available to anyone for a modest fee—and the patience to endure a long waiting list. The garden is organic, free of herbicides and pesticides, and remains primarily for the cultivation of fruits, vegetables, and herbs.

The long-awaited email finally came—after ten years on the waiting list for a garden plot in a former Victory Garden in Washington, DC, Nicholas and I got one. So in March 2018 we high-tailed it over to what is now named the Cleveland Park Community Garden and found plot B-05b, an eight-by-ten-foot rectangle that was all ours—weeds and all—for just twenty-five dollars a year.

We loved the idea that over the decades so many gardeners had worked the soil before us. Preparing the plot for our future salad bowls, we dug up old vines, bricks and planks (there once had been an apartment building on the land), faded garden markers, and worms—lots of worms.

The original Victory Garden in this spot was established in 1942 with sixteen plots, each of them about forty feet by forty feet. Peter Peart, a fellow gardener, told me about one of the original gardeners—Mrs. Sara Bell. Quite regal in appearance, she was affectionately called "Ma Bell." Already in her eighties when they met, Ma Bell proudly told Peter that she adopted a plot so she could "support the war effort." But it was equally important to her to share what she grew, and so Ma Bell often held tomato-tasting events so the greater community could enjoy her many varieties. Ma Bell continued her passionate gardening and continued to share her produce and stories well into her nineties.

When I'm working in our little plot, I like to imagine those early gardeners worrying over the latest war news, but also finding solace from their garden community. Fast forward to March 2020, when the world found itself fighting a different type of war: a pandemic. While so many of our daily activities were severely curtailed, we gardeners were lucky when the DC government permitted community gardens to operate. And so, masked and gloved, Nicholas and I returned to our plot, joyful to once again meet up with our community of gardeners, knowing we were all finding comfort in turning the soil and planting our seeds.

It's now summer 2021. The pandemic continues but we persevere—at home and in the garden. The plots are overflowing with an abundance of vegetables, and we continue the tradition of sharing our bounties with family, friends, and neighbors. It's a wonderful feeling—Mother Nature gives us a hug and we hug her back. I think Ma Bell would be thrilled to see that the Victory Garden she cultivated for more than fifty years has continued to flourish.

ANDREA PEDOLSKY
Washington, DC

Dowling Community Garden, Minneapolis, Minnesota

Two blocks west of the Mississippi River in the heart of Minneapolis, the Dowling Community Garden is located on the grounds of Dowling Elementary School, a public magnet school with a special focus on urban environmental learning. The garden was first created in 1943, when managers of public lands were encouraged to devote some of the space to Victory Gardens. Today some 250 gardeners maintain the more than 190 plots in the garden; teachers and children from the school take care of four of the plots. One section features raised-bed gardens at wheelchair level and taller raised beds for gardeners with limited mobility. In line with the broader Dowling mission, gardeners are encouraged to grow extra for food pantry donations, which volunteers collect and deliver three times a week from June through September. Organic gardening methods are encouraged; pesticides and herbicides are prohibited.

I'm a Minnesota girl born and raised, but in 2015 I moved to New York City and began working at a nonprofit called Project EATS, which works with Brooklyn schoolchildren. We taught them to grow fruits and vegetables and then set up a market to sell the produce; the students were integrated into the full process, beginning to end. That same year I entered graduate school at NYU in environmental conservation education and, as part of my studies, interned with the nonprofit Farm School NYC, which runs a comprehensive farming education program through the lens of food justice.

The concept of food abundance was something I largely took for granted while growing up in Minnesota, but this blindness changed quickly during my time in New York. In my work and studies, I came face-to-face with many injustices centered around food. I also saw and heard how those awful patterns echo throughout the country, and I wanted to find a way to use my newfound knowledge to dig out the roots of some of those injustices. That's probably why, when I came back home to Minneapolis after graduate school, I was so excited to start a job as a food systems coordinator for a major food shelf.

While planning my move back to Minnesota, I learned about the Dowling garden from friends and immediately put myself on the waiting list. Then, magically, I got a call that one spot was available, and I jumped at it. At the time I didn't know much about Victory Gardens, but that year (2018) the garden held its seventy-fifth anniversary celebration. Such a momentous occasion made the garden's history come alive for me. I could see it as a thread through time.

In my garden plot I always wanted to grow mostly flowers, partly because it's so important to provide resources for pollinators, but mostly because I just liked the idea of growing flowers and making bouquets to give away. And then George Floyd was murdered.

It was May 25, 2020. In the immediate aftermath, our city—*my* city—was ripped apart. By then, I had had my garden plot for a couple of years. My usual route from home to the garden took me through the intersection where Mr. Floyd died. After his death, that intersection was blocked off for months. But even when it was reopened, I couldn't drive that usual route. It didn't seem respectful.

During this time I realized that being part of a community garden was more important than ever. I was craving a sense of togetherness and hope, and I also knew I wanted to be involved in fostering that feeling somehow. The garden does this for me in several ways. Because it's right next to the Dowling school, the children are often somewhere in the garden, maybe working on a class project or just walking along the aisles and admiring the flowers. Neighbors walk through, sometimes with their happy dogs trailing along. Fellow gardeners trade produce and share tips. Through these moments, a network of small interactions continually spins a beautifully interconnected web.

I cherish that community. I love to stroll through the garden and stop and visit with other gardeners. The sun is shining, and there's a nice breeze, and the birds are singing, and the butterflies are doing their thing, and people are smiling—it's just magical to walk through that. And I know I'm part of a very special place.

SONJA TOSTESON
Minneapolis, Minnesota

Fenway Victory Gardens, Boston, Massachusetts

The Fenway Victory Gardens sit in one special corner of a city park in the heart of Boston. In turn that park, named the Back Bay Fens, is one of the nine individual parks linked together into an exquisitely designed system known today as the Emerald Necklace. The necklace, with its nine jewels, is largely the legacy of Frederick Law Olmsted, universally called the father of landscape architecture (a term he invented).

It started in the 1870s. Olmsted had recently completed the design of New York's Central Park, and Boston's city leaders were so impressed they asked him to design a park for Boston. Olmsted, a man of grand vision, instead suggested a complete system. In his vision, three existing gardens, together with six new ones he would design on parcels of land throughout the city, would be linked together by a curving chain of waterways and tree-lined pathways. Olmsted called it "a green ribbon" that would bring all parts of the city together.

Olmsted pushed to have the brackish marsh that Bostonians called the Fens, an area polluted with decades of sewage and "so foul that even clams and eels cannot live there," included as part of the chain of parks. It was here that the first of the six Olmsted-designed parks was developed, beginning in 1878. And it was here, in one corner of this park, that a Victory Garden was established in 1942, when the federal government began encouraging local government officials to convert some of their public lands into vegetable gardens. The oldest World War II Victory Garden in the US has served the citizens of Boston as a community garden ever since. For many of its 450 gardeners today, Olmsted's spirit is very much alive there.

Among the 450 Bostonians who have a plot in the Fenway Victory Gardens, Arthur Rose may well be the only one who actually saw military service during World War II. He has spent the last 50 of his 103 years working in this garden.

I was born in 1919 in Grand Rapids, Michigan, but we moved to North Dakota when I was two. It was a good place to grow up. We always had a garden, and my parents made me and my brother do the weeding. We both hated it; all we wanted to do was go swimming. And here I am, all these years later, in this garden. Funny how things turn out, isn't it?

I served in the navy in World War II as an engineer working with LCT [landing craft tank] flotillas. I oversaw eighteen LCTs during the Normandy Invasion, but for some reason I never felt unsafe. After the war I stayed in the naval reserves and was working in a jewelry store in Flint, Michigan, when the Korean War broke out. So I got called up and sent to a torpedo base in Newport, Rhode Island. During my time there, I fell in love with New England and as soon as I could, I moved to Boston. Then one day I happened to pass by this garden and decided gardening was something I wanted to do—finally.

Over the years, I've seen the garden improve 100 percent. It was a mess at first, on its deathbed for a while. Then a group of us decided to draw up a constitution and some rules, hoping it would stabilize things so the garden could continue. At one point some businessmen were trying to make it a parking lot for Fenway Park—the Red Sox baseball stadium is literally right across the street from the garden—but one of our state senators was a big supporter and he managed to kill that proposal.

At one point I had three plots in the garden and grew every vegetable imaginable and some wonderful raspberries. Now I just have one, and it's half flowers and half vegetables. Right now I have a beautiful crop of beets.

Back when I started with the garden it was mostly older people. Now we have a lot more younger people, and all of us have very different ideas for our gardens; some are real showplaces. There's a really nice community spirit; once a month we have work parties, and that brings people together.

Working in this garden is the highlight of my day; I look forward to coming here every day. Besides, I'm retired. What else would I do?

ARTHUR ROSE
Boston, Massachusetts

Stan Everett, a historian at heart, currently lives across the street from the garden. He has been actively involved there, as gardener and historian-observer, for more than thirty years.

I was born in 1943 and grew up in Boston, in the projects. I served three years in the army, moved out to the suburbs for a while, then came back to the city.

You probably know that the garden is part of the Emerald Necklace, designed by Frederick Law Olmsted. But you might not know that next year [2022] will be his two hundredth birthday. It seems to me that some of the folks here now don't know about him, and that's a shame.

Things change over time, of course. And some changes are for the better. One thing I notice these days is that the makeup of the gardeners is much more diverse, so that's a good thing. I've also watched changes in the garden's physical structure over the years. The original Victory Garden here was one huge plot, no borders, no fences. Even into the early 1970s, when I first had a plot there, a lot of the gardens were not fenced in. It was like a backyard retreat. I can remember lying down on the grass—you would see trees, no buildings.

One important thing hasn't changed, though—the true heart of the garden. I walk around there often, and I notice I quickly feel calm. Somehow a garden puts our minds at rest.

STAN EVERETT
Boston, Massachusetts

Wendell Booth has had his Fenway plot for more than twenty years. On the day he and I chatted by phone, he was in his garden, picking butternut squash.

My father had a hundred-acre farm in North Carolina, near Chapel Hill. I learned a lot from him. That was a long time ago, of course. Today I'm at what most people call retirement age, but I'm still working. I paint houses, the interior walls, I mean; shovel snow in the winter; and do landscaping for several people in the summertime. I get paid to exercise!

Originally I had mostly just vegetables here, but now I also plant flowers; we need the pollinators. I used to grow spinach, but with global warming it bolts too quickly, so now I grow three kinds of Swiss chard instead. I do everything from seed and save seeds for the next year wherever I can. I love to grow cherry tomatoes and snack on them while I'm working in the garden. I have a double plot here, meaning sixty by fifteen feet, so I have lots of time for snacking!

One thing I've noticed recently in this garden is that more and more people are growing vegetables. It makes sense. It's convenient, it's healthy (no pesticides), and with this COVID thing, it just seems safer. And remember, this is a public park, and the garden is open to the public. People like to come see what we're doing. College kids walk through and lots of them stop to ask questions. And people coming out of Fenway Park after a daytime game will come strolling through, asking about the garden. On the Fourth of July, we've had 300,000 people wander through here. I like to think some of them might be inspired to start a garden of their own.

At first, managing my little garden here was fun, more like a hobby. Now it's a way of life. I can pick my own vegetables when I want to, it gets me outdoors, which is safer these days, it's good exercise, plus I really like the people. And even though I live an hour away, so it takes me a total of three hours to get in an hour's work in the garden, I can't imagine not doing it.

WENDELL BOOTH
Boston, Massachusetts

Susan Povak is about to celebrate her thirtieth year nurturing her garden in this spot she calls sacred.

My garden is on sacred land. It was originally created within the Back Bay Fens, the first of six parks of the Emerald Necklace designed by Frederick Law Olmsted in 1878. Today on that site, the Fenway Victory Gardens attract visitors to view more than five hundred garden plots in this seven-acre park, offering a temporary respite from the city's frenetic pace.

While Victory Gardens membership is open to all city of Boston residents, what I love about gardening here is what it offers to *all* who visit, hailing from across the country and beyond. They marvel at how each gardener's plot is unique and holds a distinct style—from quirky to sophisticated. In addition to the garden plots, we house a bee apiary in our pollinator garden, which helps our surrounding gardens to thrive. There is a medicinal garden to teach about the healing properties of herbs. We offer workshops for our new and seasoned gardeners alike. We partner in the care of the many mature trees that form a lush canopy. And it's always fun to chat with a passing park ranger riding one of the city's mounted horses.

Next year [2022] will be my thirtieth year in the Victory Gardens. I would describe my garden as having a woodland theme with shrubs, ornamental grasses, a mature cherry tree, and meandering paths throughout. It's not your suburban backyard; we're gardening in a city, on very public land. What we create, all together, is public beauty. It's all based on a feeling of community. That's the most important thing, and that's why I'm here. That sense of community makes it all worthwhile. One of our gardeners, who has since passed away, once told me that a community gardener plants three times: once for yourself, once for your neighbor, and once for someone who needs it.

One warm summer night years back, I happened to be working in the garden while Bruce Springsteen played a concert at Boston's venerable Fenway Park, which is directly across the street from the gardens. We could clearly hear the music, and I can remember relaxing under my tree, watching the birds, while listening to the Boss singing "Thunder Road" and thinking, *It just doesn't get any better than this.*

SUSAN POVAK
Boston, Massachusetts

Why Community Gardens? A Closing Word

They're called community gardens for a reason. Not only are they physically situated in a certain community, they also, by their very nature, cultivate a sense of community among those who spend time there. Most of all, they imbue the participants with a deeper notion of the true meaning of community: the inclusive, compassionate community of all humankind.

Frederick Law Olmsted, who in the late nineteenth century created one of the gardens included here and many others besides, describes it far more eloquently than I ever could. Olmsted firmly believed that the stresses of daily life made people nervous and wary of one another, and that the calming effect of time spent in the natural world could offset those fears. Here, from his 1870 lecture "Public Parks":

> We want a ground to which people may easily go after their day's work is done, and where they may stroll for an hour, seeing, hearing and feeling nothing of the bustle and jar of the streets. . . . We want, especially, the greatest possible contrast with the restraining and confining conditions which compel us to walk circumspectly, watchfully, jealously, which compel us to look closely upon others without sympathy.

Today, more than 150 years later, the world is exponentially more complicated and life stresses more destructive, but we would do ourselves no favor if we dismiss his ideas as old-fashioned. The harder things get, the more we need to "look closely upon others [with] sympathy."

Kindness germinates in a community garden as surely as seeds, and that offers hope for us all.

ACKNOWLEDGMENTS

In Gratitude

To properly thank the many people who helped this book come into being, let us start where it all began: the sons and daughters of the original Victory Gardeners of World War II who took the time to share their stories, and their own sons and daughters who helped get the stories down on paper. My profound gratitude to you all. And to the many friends, acquaintances, and complete strangers who helped me locate those amazing storytellers: Elizabeth Draper, Portland; Victoria Larson, New York City; Kristen Mobilia, Boston; Sydney Stevens, Oysterville, Washington; Alice Stuckey, Portland; Mac Stuckey, Stuart, Florida.

For the beautiful art: Janice Yang, whose stunning color portraits grace the chapter openings (that's her on the cover), and Lee Johnston, whose wonderful pen-and-ink drawings prove once again that a clear illustration beats a thousand words every time.

For research assistance: Kathryn Rose O'Conaill, resourceful and dogged research librarian who also happens to be a very talented singer of Irish folk songs. If you ever doubt the existence of serendipity, ask me how I met her.

For wholehearted support, kind thoughts, and care packages, my immediate and extended family. (You know who you are.) And particular gratefulness to Mary Ella Kuster, the sister of my heart, who saw a way to help and would not take anything less than yes for an answer.

Thanks to literary agent Heather Jackson, for her enthusiasm about the original idea, savvy guidance, good brain, and big heart. And the smart, gracious professionals at HarperCollins who guided me through with skill and kindness.

I am also very grateful to these horticultural-industry specialists for help with information about new products and most especially for allowing me to share their beautiful photographs: Beth Wickel and Kara Zondag, Totally Tomatoes; Catherine Kaczor, Hudson Valley Seeds; Deborah Miuccio, Gardener's Supply; Diane Blazek, executive director, All-America Selections; Jennifer Tuzzeo, Botanical Interests; Joshua D'errico and Kristen Earley, Johnny's; Kristen Noble, Harris Seeds; Laura Stilson, Baker Creek; Renee Shepherd, Renee's Garden; and Roberta Weathers, Park Seed.

I wish I could seat every single one of you around my dinner table; what a conversation that would be!

Photo Gallery

From the hundreds of specific plants I've recommended earlier in this book, I've selected a particular group for this special photo gallery. I chose them based on two criteria: that I consider them especially noteworthy in some respect and that I suspect many of you may not be familiar with them.

For a rather silly example, think about radishes. Wonderful as they are in many ways, you probably don't need me to show you a picture of one; everybody knows what a radish looks like. But suppose you were reading along and came to the description of a Watermelon radish. I'm guessing you didn't even know such a thing existed—and it's pretty darned amazing. Wouldn't you like to know what it looks like? Of course you would.

As you're planning your own personal container Victory Garden, you may have that same thought often while reading about a certain plant: *I wonder what that thing looks like.* That's where this gallery comes in. The photos are listed here in the same sequence as you would encounter the plants when reading chapters 5, 6, and 7. If you don't find the exact photo you're looking for here, a simple internet search can likely help you.

Beneath many of the photos you will find the name of the seed company or organization that generously supplied them for us. You can take that to mean that they also sell those very seeds—except remember that things do change over time. If what you're interested in turns out to be no longer available, all of those fine organizations can recommend a good substitute. You might also take a look at the Resources appendix for a more detailed description of these companies and their offerings. And don't blame me if you get thoroughly swept up and miss dinner.

PHOTO GALLERY IMAGE CREDITS

A resounding shout-out to these companies and organizations for the use of their beautiful images in this photo gallery. Beneath many individual photos you will see the name of the seed company that provided it, which means they are also a great source for the item pictured. If you wish to learn more, you'll find their website address immediately below their name in the list here.

All-America Selections
 www.all-americaselections.org

Baker Creek Heirloom Seeds
 www.rareseeds.com

Gardener's Supply Company
 www.gardeners.com

Johnny's Selected Seeds
 www.johnnyseeds.com

Renee's Garden
 www.reneesgarden.com

Totally Tomatoes
 www.totallytomato.com

VEGETABLES

Asian Greens, Little Jade Chinese Cabbage

©reneesgarden.com

Asian Greens, Merlot Chinese Cabbage

Johnny's Selected Seeds

Asian Greens, Scarlette Chinese Cabbage

Shutterstock/bonchan

Asian Greens, Mizuna

Shutterstock/ElenVik

VEGETABLES

Asian Greens, Central Red Mizuna

Johnny's Selected Seeds

Asian Greens, Green Fortune Pak Choi

©reneesgarden.com

Asian Greens, Tatsoi

Shutterstock/meeboonstudio

French Mascotte Beans

©reneesgarden.com

Royal Burgundy Beans

Shutterstock/Master the moment

Formanova Beets

Baker Creek Heirloom Seeds/rareseed.com

Chioggia Beets

Shutterstock/Art_Pictures

Golden Beets

Shutterstock/vaivirga

VEGETABLES

Parisienne Carrots

Baker Creek Heirloom Seeds/rareseed.com

Romeo Carrots

©reneesgarden.com

New Kuroda Carrots

Baker Creek Heirloom Seeds/rareseed.com

Leafy Greens, Bright Lights Swiss Chard

Johnny's Selected Seeds

Leafy Greens, Green Curls Kale

©reneesgarden.com

Leafy Greens, Red Giant Mustard Green

Shutterstock/Peter Turner Photography

Lollo Rossa Lettuce

Shutterstock/Chatcham172

Tom Thumb Butterhead Lettuce

Baker Creek Heirloom Seeds/rareseed.com

VEGETABLES

Garden Babies Butterhead Lettuce

©reneesgarden.com

Little Gem Romaine Lettuce

Baker Creek Heirloom Seeds/rareseed.com

Snak Hero Peas

All-America Selections

Little Crunch Peas

©reneesgarden.com

Pink Beauty Radish

©reneesgarden.com

Easter Basket Radish mix

Baker Creek Heirloom Seeds/rareseed.com

French Breakfast Radish

Shutterstock/Skrypnykov Dmytro

Watermelon Radish

Shutterstock/Sacha van der Veen

VEGETABLES

Yellow Patio Choice Tomato

All-America Selections

Tidy Treats Tomato

Johnny's Selected Seeds

Lizzano Tomato

All-America Selections

Litt'l Bites Tomato

Gardener's Supply

Sungold Tomato

Shutterstock/Jess Abeita

Ruby Crush Tomato

Totally Tomatoes and Sakata Seed America

Celano Tomato

All-America Selections

Sunrise Sauce Tomato

Johnny's Selected Seeds

HERBS

Purple Basil

Shutterstock/emrahh

Siam Queen Basil

All-America Selections

Windowbox Mini Basil

©reneesgarden.com

Persian Basil

All-America Selections

Garlic Chives

Shutterstock/ahmydaria

Golden Oregano

Shutterstock/OA2

Rosemary

Shutterstock/janaph

Lemon Thyme

Shutterstock/TAGSTOCK1

EDIBLE FLOWERS

Tuberous Begonias, assorted colors

Shutterstock/Dorothy Chiron

Dianthus mixture

Shutterstock/EQRoy

Nasturtiums, trailing style

Shutterstock/Edita Medeina

Jewel Mix Nasturtiums

Shutterstock/Jack N. Mohr

Alaska Nasturtiums

Shutterstock/Julie Vader

Gleam Nasturtiums

Shutterstock/mizy

Nasturtium, Cherries Jubilee

©reneesgarden.com

Pansies, assorted colors

Shutterstock/Jerry Gantar

APPENDIX: RESOURCES FOR GARDENERS

Mail-Order and Internet Sources

First, I need to point out a few things.

1. This list is in no way comprehensive. This is especially true of the e-commerce suggestions. The ones listed here are not the only possibilities, not by a long shot. They are simply the ones I am familiar with, admire, and gladly recommend to you. A highly opinionated list of favorites, you might say.
2. The companies and organizations listed here are separated into just two large groups: (1) horticultural industry and gardening organizations, and (2) e-commerce companies, those wonderful folks who offer a range of garden products by mail order. Companies in that second, much larger category are listed alphabetically, and I urge you to read all the descriptions carefully, to uncover their main attributes. Some companies specialize; some are all-purpose. Some are geared to a particular climate range. Some sell plants as well as seeds. Some provide a wide range of gardening tools and equipment, either in addition to seeds or sometimes as their main focus. Some offer their catalogs online only; some have both print and online (and all of them are bursting with helpful information).
3. The information listed here is accurate as of mid-2022, but as you know, things change—often for the better—in this particular marketplace. I know you recognize

that, so this is just a gentle reminder for patience if you run into a glitch here or there. And if you find a new favorite, I'd love to hear about it.

4. As you explore the websites of the seed houses here, you will often be presented with an offer to subscribe to their newsletters. My enthusiastic advice is—accept. They're almost always free, and always helpful, and easy to unsubscribe from, if it doesn't suit. Even the brief messages that pop up in your email inbox are useful: headlines like "Ready to plant tomatoes?" can serve as a good kick in the blue jeans for those times when the calendar gets away from us.
5. A special note about ordering live plants by mail: common sense tells us that traveling long distances is hard on defenseless little plants, so try to find sources as close to you as possible. That's one reason I scouted around for companies based in various regions. All these suppliers take great care to package well, but if you let the box sit unattended for days after it arrives, you're just asking for disappointment.

Organizations: Information, Inspiration, and Commiseration

ALL-AMERICA SELECTIONS (AAS)

https://all-americaselections.org

From their website: All-America Selections is an independent nonprofit organization that tests new, never-before-sold varieties for the home gardener. After a full season of anonymous trialing by volunteer horticulture professionals, only the top garden performers are given the AAS Winner award designation for their superior performance. AAS is the only national, nonprofit plant trialing organization in North America.

From me: Every AAS winner has its own page on the website, with one or more photos, growing instructions, and, in more recent years, notes from the judges about features they especially loved. We gardeners owe a great debt to these volunteer growers and judges. The results of their dedicated work are so well respected that many seed houses set aside a special highlighted section on their websites for AAS winners; they know their customers will not be disappointed.

NATIONAL GARDENING ASSOCIATION

https://garden.org

From their website: Since 1971, our mission is to promote gardening. With over a million members, we are the largest social media website dedicated exclusively to gardening. We teach people how to get started gardening and improve their plant-growing skills.

From me: And how! This is an incredible resource, free to all users (you don't have to be a member to access the info), absolutely crammed with information yet wonderfully easy to use. For example, with just your zip code you can find your USDA zone, your local frost dates, and a comprehensive planting calendar that matches your climate patterns. Another example: from their jampacked website, one (just one of many) main category is Garden Learning Library, where one (just one) subcategory is Food Gardening Guides, which is further subdivided into Vegetables, Fruits, and Herbs; in turn, the Vegetables group offers more than 220 articles that provide comprehensive information for eleven groups of most popular vegetables, further broken down into in four subheadings: Getting Started, Planting, Care, and Harvesting. Whew.

NATIONAL GARDEN CLUBS, INC.

https://gardenclub.org

If you're interested in joining a local garden club but don't know anyone who can point you in the right direction, start with the home page of this national organization. At the top, find the tab called "Join a Club," be sure to allow the site to access your location, then enter your zip code at "Find a Club." Read the descriptions and choose one that fits you best, either for location or special interest.

COUNTY EXTENSION OFFICES

Extension offices, under the umbrella of the US Department of Agriculture, are rich with resources for home gardeners. The easiest way to access the one you want is to simply ask your favorite search engine for "extension office for [XYZ] county, state."

Mail-Order Sources: Seeds, Plants, Tools, Equipment, and Good Advice

BAKER CREEK HEIRLOOM SEEDS

2278 Baker Creek Rd., Mansfield, MO 65704
www.rareseeds.com (for online catalog and orders)
seeds@rareseeds.com (for questions)
Catalog: Print and online

A family-owned seed company with a passion: "saving our agricultural and culinary heritage." That means they sell only heirloom seeds and strive to find rare varieties from Asia and Europe, including many that date from the nineteenth century. Sounds like their offerings would be limited, doesn't it? Nope. They have seeds for well over one thousand vegetables, herbs, and flowers. Some are unusual and exotic, some are cozily familiar, and to my mind every single one is intriguing. The catalog is filled with gorgeous photography, and shipping for seeds is free. Wonderful staff, seeds you won't find anywhere else, and did I say free shipping? This company is an absolute gem.

BOTANICAL INTERESTS, INC.

660 Compton St., Broomfield, CO 80020
www.botanicalinterests.com
877-821-4340 (customer service)
customerservice@botanicalinterests.com
horticulturist@botanicalinterests.com (for gardening questions)
Catalog: Print and online

Botanical Interests is a small, family-owned seed company that has won my heart with its operating philosophy: focus on products that align with principles of organic gardening and sustainability, give back to the community through seed donations and fundraising support, and most of all, make the seed packets as useful as humanly possible—and then also make them beautiful. The results are stunning: exquisite artwork on the front, essential planting info on the back, and inside there is a complete foldout guide crammed with directions on when to harvest, recipes, history, easy growing tips, organic pest control, and more, all specific to that plant. So if by the time your seeds arrive you can't find the catalog you ordered from, everything you need to know is right at hand.

This same "be helpful" mandate is also apparent on the website. They offer a range of free, inspiring gardening information that is nothing short of astonishing. Blogs with recipes, growing guides searchable by specific plant, a glossary of terms, articles about every gardening technique and situation you could ever encounter. Here's just one example out of a bajillion: In the main Articles tab, there's a category called Edibles, and nestled there is an article titled "Edibles for Partial Shade." After a general intro, this compact article then lists and links to ten leafy greens and ten herbs that can survive with as little as two hours of sun, and seven root crops, ten other vegetables, and thirteen herbs that can manage with just four sunny hours. Not all are ideal for containers, but that's where your wish list comes in.

Also very helpful are the curated seed collections—Microgreens, Baby Greens sampler, Basic Bounty vegetables (one herb and eight vegetables, most of which would do well in containers), and Container Vegetable collection, with container-appropriate varieties of ten popular veggies.

GARDENER'S SUPPLY

128 Intervale Rd., Burlington, VT 05401
www.gardeners.com
800-876-5520
Orders: 888-833-1412
Catalog: Print and online

With its strong focus on the "hardware" side of gardening, this employee-owned company is unique on our list but well worth your time to explore. Their catalog is a mind-boggling array of garden products: tools (many of them developed by in-house designers), supplies, raised beds, pots, planters (including some very handsome grow bags), supports, even some seeds and plants. Just as impressive is the amazingly broad offering of online education. Under the Advice tab on the home page, you will find scores of how-to articles, planning guides, and so much more. And if you don't find just what you need there, you can use the online submission form to "Ask a Gardener" about anything garden related. The company has four proprietary retail stores, all in New England, and some of their products are carried in garden centers and home improvement stores nationwide.

GROWERS EXCHANGE

951 Techpark Place, Sandston, VA 23150
www.thegrowers-exchange.com
888-829-6201
Catalog: Online only

This company sells only herbs, and only as live plants, no seeds. The selection is large, and most plants are priced at between $8 and $10; shipping starts at $8.95.

GROWJOY

www.growjoy.com
Catalog: Online only

A small, family-run operation that offers only plants, no seeds, all grown in their own greenhouses in Berne, Indiana. Navigate your way through their full selection of vegetables and flowers by exploring the grouping called "Plants for Productive Patios" under the "Shop by Purpose" subhead on their home page. Look for items that correspond to your personal wish list. Most of the plants are priced (in 2022) at $6.95; orders more than $50 ship free, otherwise a flat rate of $14.99.

GURNEY'S SEED & NURSERY CO.

P.O. Box 4178, Greendale, IN 47025
www.gurneys.com
Phone orders: 513-354-1491
Customer Service: service@gurneys.com; 513-354-1492
Catalog: Print and online

Dating from 1866, this major garden company has been well-known to home gardeners for generations. Today its catalog features just about every kind of plant you could imagine: vegetables, herbs, fruits, flowers, perennials, shrubs, trees, as well as a good selection of tools and supplies. Of particular interest for us, a number of their vegetables are available as both seeds and live plants. With some experimenting, the best approach I found to identify these choices is this: ask the search bar for "vegetable plants" and quickly scan the results grid (it includes herb plants along with veggies). If there is a wide range of prices given for a certain item (such as $3.99 to $24.99), you'll know they sell plants of that item as well as seeds. But you'll also need your own wish list to zero in on varieties that work nicely in containers.

HARRIS SEEDS

355 Paul Rd., Rochester, NY 14624
www.harrisseeds.com
800-544-7938
Catalog: Print and online

Since 1879, Harris has provided both home gardeners and commercial growers with vegetable and flower seeds and a wide range of tools and supplies for garden and greenhouse. Today's container gardeners will find lots to choose from in the grouping labeled "container gardens" under the main Vegetable tab on the home page.

HUDSON VALLEY SEED COMPANY

4737 Route 209, Accord, NY 12404
www.hudsonvalleyseed.com
845-204-8769
mail@hudsonvalleyseed.com

From their website: We believe that a seed is more than meets the eye: it is a time capsule telling tales of the plants, crops, and people that came before us, and we work with a range of artists as diverse as our seeds to pass those stories down to present and future gardeners.

From me: That lovely sentence embraces the two pillars upon which this company proudly—and successfully—stands. First, a dedication to principles of sustainability, organic methods, and preserving seed diversity. Second, and every bit as important, a commitment to supporting art and storytelling. It's not your typical horticultural business plan, but it works beautifully.

For the seed part of the business, they carry a wide selection of seeds for vegetables, herbs, and flowers, including many edible flowers. All their seeds are open-pollinated (which means pollinated by natural forces, as opposed to controlled cross-pollination performed intentionally to produce new cultivars) and most are organic and heirloom. And all of this is supported with a generous range of garden tips and planting guides.

It's the art part that warms my heart. That ideal is put into practice with special seed packets they call Art Packs. Each one has an outer fold-open packet featuring original art on the front and a story about the plant and an introduction to the artist on the inside, then an inner packet with the seeds and planting directions. Each year, ten to twelve new plant introductions are illustrated by ten to twelve new artists, so that by now most of the catalog offerings are available in either Art Packs or standard packets. The cost differential is about 60 cents—a small price, in my opinion, for such charm.

JOHNNY'S SELECTED SEEDS

13 Upper Main Street, Fairfield, ME 04937
www.johnnyseeds.com
877-564-6697
Catalog: Print and online

Johnny's celebrates its fiftieth anniversary in 2023, a testament to the vision of founder Rob Johnston, who started off in a New Hampshire farmhouse attic at the age of twenty-two. Today, he and his wife still shepherd the company, now with one hundred–plus employees who are also owners of the business. From the beginning, Rob intended to offer seeds of unusual varieties, even when it meant producing the seeds himself. Today, Johnny's is widely admired for its extensive plant-breeding program, which over the years has introduced home gardeners to many new favorites, including numerous All-America Selections.

The easy-to-navigate catalog has well over a thousand varieties of vegetables, herbs, flowers, and fruits, plus an astonishing collection of garden tools and supplies, with everything you could ever need and some things you didn't even know existed. One caution: Johnny's supplies large-scale market gardeners, too, so be sure you're looking at a package size or quantity that fits your small-garden space. And I am completely blown away by the breadth of gardening information on their website. From the home page, open up the Grower's Library tab and prepare to be amazed. If you cannot quickly find something or have a problem, try their "Ask a Grower" feature; a real live person will respond.

KITIZAWA SEED COMPANY

201 Fourth St., #206, Oakland, CA 94607
www.kitazawaseed.com
seeds@kitizawaseed.com
510-595-1188
Catalog: Print and online

Founded by two Japanese brothers in 1917, Kitazawa is today the premier US source for seeds of Asian vegetables. Recently, in response to customer requests, they have also added varieties of Mexican vegetables, and all of them packaged in the distinctive Kitazawa style: simple manila envelopes with dark-green ink. The catalog is the same: no fancy graphics, just clear information about their many varieties, including several pages of recipes. I really admire their curated collections, called Chef Specialty Gardens. Currently, there are nineteen collections, with themes such as Baby Leaf Garden, Thai Garden, and Edible Flower Garden. Each one has seven different packets, a thoughtful mix of familiar and new-to-you varieties.

NICHOLS GARDEN NURSERY

1136 Main St., Philomath, OR 97370
PO Box 1299, Philomath, OR 97370
www.nicholsgardennursery.com
800-422-3985 (phone orders)
Catalog: Print and online

This well-respected nursery, founded in 1950, is family-owned and operated, now into the third generation. Known for high-quality seeds for vegetables, flowers, and herbs, along with garden tools and supplies.

PARK SEED

1 Parkton Ave., Greenwood, SC 29647
www.parkseed.com
info@parkseed.com (July and August only)

One of America's oldest and largest mail-order seed companies, Park dates from 1868, when fifteen-year-old George W. Park decided to earn some spending money by selling seeds from his own garden. At 154 years old and still going strong, the company provides vegetable gardeners the comfort of longtime favorites and the joy of new treats. The 2022 online catalog has seeds for 119 vegetable varieties that are appropriate for containers, including some very appealing mixes and blends. Put "container vegetable plants" in the search box. Also on the website, along with an impressive array of tools and supplies, is a rich selection of gardening articles, planting guides, tip sheets, and even recipes.

RENEE'S GARDEN SEEDS

6060 Graham Hill Rd., Felton, CA 95018
www.reneesgarden.com
888-880-7228
Catalog: Online only

Renee Shepherd, like several others included in this section, is one of my heroes. She is fiercely dedicated to offering gardeners the very best seeds, only the ones that deliver, as she puts it, "great flavor, easy culture, and exceptional garden performance." When you explore her website, which I hope you will, look for the essay titled "How We Choose Our Seeds." It's an illuminating glimpse into the challenges all seed companies face as well as an inspiring personal manifesto.

But I have to warn you: being on the website can make you feel like you just walked into the world's biggest candy store and have no idea where to start. This is especially true for container gardeners, because as the website proudly exclaims, "Space-saving varieties are our specialty!" How you navigate through depends on your shopping style. If you're that person who needs to look at every single style, shape, fabric, and color before deciding on a pair of socks, use this: reneesgarden.com/collections/great-in-containers.

If you prefer a more focused approach, choose "Container Gardening" on the drop-down menu under Gardening Resources. That gives you a rich array of articles and instructional videos on various aspects of container gardening, but I suggest you start with the first item: "container specialty varieties." It's a comprehensive list of names of, and links to, seventy-four vegetables and herbs especially chosen for container-garden success; have your wish list handy. Even simpler, if you just want to get the darn socks *right now*, take a look at these two special collections, each containing five seed packets plus a planting guide: the Container Kitchen Garden collection (tomato, lettuce, chard, carrot, basil) and the Container Herb Garden collection (cilantro, basil, chives, dill, parsley).

SOUTHERN EXPOSURE SEED EXCHANGE

P.O. Box 460, Mineral, VA 23117
www.southernexposure.com
gardens@southernexposure.com
540-894-9480
Catalog: Print and online

As you might deduce from the name, this company specializes in varieties that perform best in the Mid-Atlantic and Southeast. I sent a catalog to my brother in North Carolina a few years ago, and now he buys nothing else. But lots of their offerings are the garden standards that *all* gardeners rely on, regardless of their zone, and even for many of the plants we usually associate with hot climates, they often include varieties with early maturities, making them a reasonable bet for cooler regions. The vast majority of their seeds are organic, and many are also heirlooms. The website features a mind-boggling collection of Growing Guides: sixteen general and season-specific guides, nine guides on various aspects of seed saving, and sixty separate guides for individual vegetables.

TERRITORIAL SEED COMPANY

PO Box 158, Cottage Grove, OR 97424
www.territorialseed.com
541-942-9547 (customer service and gardening questions)
800-626-0866 (phone orders)
Catalog: Print and online

This family-owned company was founded in the Pacific Northwest by Steve Solomon, something of a legend among home gardeners who strive toward self-sufficiency. It has been owned since 1985 by Tom and Julie Johns, who, along with their seventy employees, remain dedicated to Steve's mission of helping gardeners produce an abundance of fresh-from-the-garden food. They offer both seeds and live plants, all of them trialed and grown at their own farms.

The selection of transplants is impressively broad, and a very respectable number of them are appropriate for containers. Put "container vegetable plants" into the home page search box, then use the menu on the left side to navigate to "product type" and choose "transplants." The 2022 catalog shows a total of thirty-nine vegetables that work in containers, and eighteen of them are available as transplants as well as seeds; that's a darned good start. On the website you'll find terrific information in the downloadable Growing Guides, with two general guides for seasonal growing and then forty-four individual guides for specific veggies.

TOTALLY TOMATOES

334 W. Stroud St., Randolph, WI 53956
www.TotallyTomato.com
800-345-5977
Catalog: Print and online

It's not *totally* in the sense of "all we do is tomatoes"; it's more like a teenager's imprimatur of awesomeness. Because their tomato selection is truly awesome, but they offer a lot more besides. From the 2022 catalog: seeds for 91 peppers, grouped into 5 categories from mild to superhot; seeds for 21 other types of vegetables, from broccoli to zucchini; and oh yes, seeds for 396 tomato varieties, in 17 distinct categories, 2 of which are especially useful for container gardeners: Patio/Container Types (34 varieties) and Hanging Baskets (5 varieties). And that, my friends, is a lot of tomatoes. Altogether, there are more than twenty tomato varieties available as live plants, including two of the Roma type. Also plants for peppers, eggplants, and onions, if you're ready to expand your repertoire. And their selection of garden tools and supplies is especially well thought out.

Index

A

All-America Selections (AAS), 36, 234
Allen, Dorothy Dodd, 115
ants, 202
aphids, 86–87, 190, 194
Armistice Day, 3
Asian greens, 38, 91–95, 219–220

B

baby plants, seeds *versus,* 38–40
Back Bay Fens, Boston, Massachusetts, 209–213
Baker Creek Heirloom Seeds, 236
baker's rack, 61–62
Bankey, Louise, 59
basil, 22, 31, 38, 53, 77, 149–154, 162, 170, 228, 243
beans, 11, 12, 13, 22, 31, 33, 38, 39, 42, 44, 71, 74, 96–98, 99, 119, 123, 220, 221
bees, 71, 151, 156, 189, 202
beets, 13, 15, 33, 35, 38, 39, 71, 75, 100–102, 123, 210, 221
begonias, 38, 180–184, 230
Bell, Sara, 206
bench, dual-purpose, 65–66
Benner, Marti, 26
birds, 10, 65, 123, 189, 202, 208, 213
blossom end rot, 87, 136
bok choi/pak choi, 95, 220
bolting, 112–114
Booth, Wendell, 211–212
Boston, Massachusetts, Fenway Victory Gardens, 209–213
Botanical Interests, Inc., 236–237
boxes, repurposing of, 66–67
Bright Lights Swiss chard, 108, 109, 222
Briley, Suzanne, 12
bucket, uses, 53
Burbank, Luther, 2
bush beans, 22, 33, 96
butterflies, 71, 156, 189, 202, 208

C

cabbage, 33, 35, 93-94, 99, 110, 111, 137, 219
carrots, 11, 13, 22, 25, 31, 33, 35, 38, 39, 64, 71, 82, 99 103–105, 115, 130, 137, 168, 188, 222
ceramic containers, 49. *See also* containers
chard, 25, 82, 91, 107–109, 116, 211, 222. *See also* leafy greens
Chase, Sara, 123
cherry tomatoes, 31, 33, 132, 140–142. *See also* tomatoes
chicken wire, 120, 121, 134
Chinese cabbage, 93–94, 219
Chioggia beet, 35, 102, 221
chives, 155–157, 229
chopsticks, 52, 53, 65, 74
clamshell-style containers, 52, 53, 75
Clark, Marilyn, 42

clay containers, 49. *See also* containers
Cleveland Park Community Garden, Washington, DC, 206–207
clutter control, 65–67, 68, 69
cocktail size tomatoes, 129, 140, 142. *See also* tomatoes
community gardens
  Cleveland Park Community Garden, Washington, DC, 206–207
  Dowling Community Garden, Minneapolis, Minnesota, 207–208
  Fenway Victory Gardens, Boston, Massachusetts, 209–213
  locations of, 205
  purpose of, 213–214
compact growth of plants, 77
compost, 58
container collections, 36–37
container gardening
  overview of, xi
  planning for, 23–24, 44
  process of, 22–23
  resources for, 40–41, 43
  timing for, 44
containers
  anchoring, 28, 30
  baker's rack for, 61–62
  choosing, 28, 48–49
  clay, 49
  cord for, 28, 30
  creativity regarding, 50
  filling, 79–81
  golden rule of, 49
  moving, 30
  multiple plantings within, 62–64
  overview of, 48–50
  painting, 68
  pyramids of, 60–61
  size measurements of, 48–49
county extension office, 18, 29, 41, 43, 99, 178, 235
COVID-19 pandemic, 16–18, 212
cultivator, 51
cut and come again process, 94
cutworms, 86, 137

D

daylilies, 33
deadheading, 53
dianthus, 38, 185–187, 230
direct sowing, 43, 71, 72
diseases, 86–87
dishpan, 53
Dowling Community Garden, Minneapolis, Minnesota, 207–208
drainage, container, 28, 30, 49, 50, 80
Draper, Nancy, 11
dual-purpose bench, 65–66
Duddridge, Norman, 9–10

E

Emerald Necklace, Boston, Massachusetts, 209–213
English peas, 118, 122, 124. *See also* peas
Europe, 2–14
Everett, Stan, 211

F

fabric bag containers, 49–50. *See also* containers
farmers market, buying plants from, 43, 77
Fech, George, 9–10
Fech, John, 9–10, 41
Fech, Viola, 9
Fenway Victory Gardens, Boston, Massachusetts, 209–213
fertilizer/fertilizing. *See also specific plants*
  overview of, 56–60
  personal story regarding, 59
  preparation of, 79
  process of, 85
  replacement of, 85
  for tomato plants, 136
  usage of, 85
flowers, edible
  begonias, 38, 180–184, 230
  cooking of, 184, 187, 193–194, 199–200
  dianthus, 38, 185–187, 230
  golden rule for, 177
  nasturtiums, 36, 38, 39, 71, 189–194, 230, 231
  overview of, 177–178
  pansies, 197–200, 231
  safety regarding, 178–179
  step-by-step preparation of, 179
  tuberous begonias, 180, 230
  violas, 22, 197–200
flowers, of basil, 151

flowers, of chives, 156–157
Floyd, George, 208
food, injustices regarding, 207–208
food-to-plant ratio, 32–33
Formanova beet, 102, 221
Fraikor, Fred, 27
France, World War II and, 7
frost dates, 29
Funk, Mary, 166

G

Gardener's Catch-22, 39
Gardener's Supply, 237–238
garden group/club, 40–41, 76–77
gardening, benefits of, 21–22
garden peas, 118, 122, 124. *See also* peas
garlic chives, 157, 229
Germany, 3, 7–8, 14, 196
globe basil, 153–154
gloves, 51, 52, 79, 86–87
golden beet, 102, 221
golden oregano, 160, 229
green beans. *See* beans
green garlic, 64–65
green peas, 118, 122. *See also* peas
Growers Exchange, 238
Growjoy, 238
Gurney's Seed & Nursery Co., 238–239

H

hand pruner, 51
hardening off, 74
hardware, 47–54. *See also* containers
Harris Seeds, 239
herbaceous perennial, 171
herbs
  basil, 22, 38, 149–154, 228
  chives, 155–157, 229
  Container Herb Garden collection, 243
  good ratio of, 33
  growing seasons of, 38
  marjoram and oregano, 158–160, 229
  Mediterranean, 148
  mint, 161–163
  overview of, 147 –148
  parsley, 38, 164–166
  rosemary, 167–169, 229
  tarragon, 170–172
  thyme, 173–175, 229
Hitler, 7
hogwire, 134
household items, as gardening tools, 52–54
Hudson Valley Seed Company, 239–240

I

influenza pandemic, 3–5
inorganic fertilizer, 57–58. *See also* fertilizer/fertilizing
Italian basils, 152
Italy, 8

J

Japan, 14
Johnny's Selected Seeds, 94, 109, 129, 141, 152, 218, 219, 220, 222, 226, 227 240–241

K

kale, 109–110, 223. *See also* leafy greens
Kitizawa Seed Company, 241
knife, 64, 93, 94, 107, 109, 156, 171

L

leafy greens, 33, 38, 106–111, 222, 223
Lee, Jim and Marian, 6–7
lemon thyme, 175, 229
lettuce, 22, 37, 38, 112–116, 223, 224
Liberty Loan parade, 4–5

M

Mackey, Cora, 123
magic ratio, 32
mail-order sources, 43, 236–244
maintenance, overview of, 84–86. *See also specific plants*
manicure scissors, 53
marionberries, 99
marjoram and oregano, 158–160, 229
Marler, Jerri Austin, 61
Master Gardener program, 41, 43, 76–77
Mediterranean herbs, 148. *See also* herbs
mesclun, 116

mildew, 87
milk jugs, 53–54, 80–81
Minneapolis, Minnesota, Dowling Community Garden, 207–208
mint, 152, 161–163, 172, 199
mizuna, 35, 94–95, 116, 219, 220. *See also* Asian greens
mustard, 35, 91, 106, 109–110, 116, 194, 223. *See also* leafy greens

N

nasturtiums, 38, 39, 189–193, 230, 231
National Garden Clubs, Inc, 235
National Gardening Association, 235
National War Garden Commission, 2–3
Nichols, Clayton, 34
Nichols Garden Nursery, 241
nitrogen, 55, 56–57, 92, 93, 107, 112, 136, 150, 199
"no bare dirt" rule, 62–64, 130
Novak, Dorothy Reinking, 139
nursery, as resource, 41
nutrients, in fertilizer, 56–57

O

Olmsted, Frederick Law, 209, 211, 212, 213–214
onion chives, 156–157
online catalogues, 35
oregano and marjoram, 35, 158–160, 229
organic fertilizer, 57–58. *See also* fertilizer/fertilizing
organizational sources, 234–235

P

Pack, Charles Lathrop, 2–3, 5–6
painting, for visual clutter control, 68
pak choi/bok choi, 95, 220
pansies, 197–200, 231
Pape, Pat, 54
Park Seed, 242
parsley, 38, 164–166
Pearl Harbor, attack on, 8
Peart, Peter, 206
peas, 33, 38, 117, 124–130, 224
peat pellets, 73–75
Pedolsky, Andrea, 206–207
pests, 86–87
phosphorus, 56, 57
pinching, 150, 151
planning, 23–24, 44
plants. *See also specific plants*
  aesthetics of, 35
  bad ratio of, 33
  characteristics of, 31–35
  checklist for, 31
  choosing, 30–37
  choosing for beauty, 78–79
  compact growth of, 77
  as container appropriate, 31–32
  food-to-plant ratio of, 32–33
  good ratio of, 33
  list for, 30–31
  magic ratio for, 32
  as not in the supermarket, 35
  researching of, 35–37
  seasonal, 37–38
  seeds *versus*, 38–40
  supply and demand regarding, 32
  what to look for, 77–78
  where to buy, 76–77
  as worth the trouble, 33, 35
plastic containers, 49. *See also* containers
Poland, 7–8
pollen, 202
pollination, 202–203
Ponte, Mary (Dee), 196
potassium, 56
potatoes, 7, 33
potting soil, 55–56, 79
Povak, Susan, 212–213
Project EATS, 207–208
pruning, 53, 150
"Public Parks" (Olmsted), 213–214

R

radishes, 22, 33, 38, 64, 71, 104, 126–130, 225
railroad, 7
Reinking, William, 139
Renee's Garden Seeds, 242–243
resources, for container gardening, 40–41, 43
rhubarb, 33
rocks, drainage and, 80
Roffler, Ralph, 99
Roma tomatoes, 143. *See also* tomatoes

Roosevelt, Eleanor, 10–12, 13
Roosevelt, Franklin D., 8
root-bound plants, 78, 79
Rose, Arthur, 210
rosemary, 167–169, 229

S

Sanborn, Bill, 137
seasons, plants within, 37–38. *See also specific plants*
seeds
  baby plants *versus,* 38–40
  development of, 202
  direct sowing of, 71, 72
  overview of, 71–72
  peat pellets for, 73–75
  saving, 72
  starting indoors with, 72–73
  thinning, 75–76
shade, planning for, 24–25
shell shock, 3
six-pack, 81
slow-to-bolt plants, 37
slugs, 86, 112–113
snap beans. *See* beans
snap peas, 118, 124–125. *See also* peas
snow peas, 118, 122, 124. *See also* peas
software (soil and additives), 55–56, 58, 60. *See also* fertilizer/fertilizing
soil, potting, 55–56, 79
Southern Exposure Seed Exchange, 243
space, for container gardening
  baker's rack for, 61–62
  container pyramids, 60–61
  green garlic and, 64–65
  making the most of, 60–65
  shade and, 24–25
  succession planting for, 62–64
  sunlight and, 24–25
  trellises for, 62, 120–121, 191
  water availability and, 25
  wind and, 25, 28
spinach, 111. *See also* leafy greens
spring lettuce, 116
Staudenraus, Margaret, 13
Stein, Gertrude, 144
Stevens, Nyel, 188
Stonebreaker, Sandy, 82
storage systems, 65–67
string beans. *See* beans
succession planting, 23, 62–64
sugar snap peas, 63, 118, 122, 124–125. *See also* peas
sunlight, 24–25, 29
supports, pea, 119–121
supports, tomato, 133–134
Sweet basil, 152
sweet peas, 118

T

tarragon, 170–172
tatsoi, 95, 220
tender perennial, 158
Territorial Seed Company, 243–244
Thai basil, 153
thinning, seeds, 75–76
thyme, 173–175, 229
Tinker Field, 34
tomato cage, 134
tomatoes, 132, 133–138, 140–142, 226–227
Tom Thumb Butterhead lettuce, 116, 223
tools, 51–54. *See also specific tools*
topsoil, 56
tornado, 34
Tosteson, Sonja, 207–208
Totally Tomatoes, 244
transplants, seeds *versus,* 38–40
tree branch, as tomato support, 133–134
trellises, 62, 120–121, 191
Tripartite Pact, 8
trowel, 51
tuber, defined, 180

U

US Department of Agriculture, 28–29, 41

V

vegetables
  Asian greens, 91–95, 219, 220
  beans, 22, 33, 38, 96–98, 220, 221
  beets, 33, 38, 100–102, 221
  bok choi/pak choi, 95

cabbage, 33
carrots, 22, 38, 103–105, 222
chard, 107–109
guidelines for, 89–90
kale, 109–110
leafy greens, 33, 38, 106–111, 222, 223
lettuce, 22, 37, 38, 112–114, 116, 223, 224
mesclun, 116
mizuna, 94–95, 219, 220
mustard, 110–111, 223
peas, 33, 38, 117–125, 224
photos of, 219–227
potatoes, 7
radishes, 22, 33, 38, 104, 126–130, 225
spinach, 111
tatsoi, 95, 220
tomatoes, 33, 131–138, 140–144, 226–227
Victory Gardens
advertisement for, 2, 15
during COVID-19 pandemic, 18
history of, xii, 1
origin of, 5–6
personal story regarding, 9–10, 11, 12, 13, 26, 27, 34, 42, 54, 59, 82, 83, 99, 115, 123, 137, 139, 166, 188, 196, 206–208, 209–213
popularity of, 12–13
purpose of, 9
revival of, xi
statistics of, 3, 14
violas, 22, 38, 197–200

W

Ward, Kay, 83
War Garden Victorious, The (Pack), 6
Washington, DC, Cleveland Park Community Garden, 206–207
water gels, 58, 60, 79
watering can, 52
watermelon radish, 129, 225
water/watering, 25, 82, 84–85, 136
WAVES, 7
weather, 24–25, 28–29, 34, 37–38, 44, 63, 85
websites
All-America Selections (AAS), 234
Baker Creek Heirloom Seeds, 236
Botanical Interests, Inc., 236–237
Gardener's Supply, 237–238
Growers Exchange, 238
Growjoy, 238
Gurney's Seed & Nursery Co., 238–239
Harris Seeds, 239
Hudson Valley Seed Company, 239–240
Johnny's Selected Seeds, 109, 240–241
Kitizawa Seed Company, 241
National Garden Clubs, Inc., 235
National Gardening Association, 235
Nichols Garden Nursery, 241
Park Seed, 242
Renee's Garden Seeds, 242–243
Southern Exposure Seed Exchange, 243
Territorial Seed Company, 243–244
Totally Tomatoes, 244
weeding, 85–86
Wickard, Claude, 10–12, 13
Wiegardt, Dobby, 99
wind, 25, 28
windowbox mini basil, 154, 228
wire cage, 134
wood containers, 49. *See also* containers
World War I, 1, 2, 3, 5
World War II, xii, 1, 7–8, 9, 10, 14

Y

Yellow Patio Choice tomato, 226

BASIL
SALAD

About the Author

MAGGIE STUCKEY is a gardener who cooks, a cook who gardens, and a writer happily immersed in both arenas, as the titles of her books suggest: *Gardening from the Ground Up, Soup Night, Country Tea Parties, The Complete Spice Book, The Complete Herb Book, The Houseplant Encyclopedia*, and five others. She divides her time between Portland, Oregon, and the tiny coastal town of Ocean Park, Washington.